Studies in Print Culture and the History of the Book

D1036538

What a Book Can Do

"Just say the blow was inflicted by a blunt instrument."
By Gordon Brooks, © 1963 *Yankee Magazine*, used with permission.

What

The Publication

a Book

and Reception of

Can Do

Silent Spring

~~~~~~

Priscilla Coit Murphy

University of Massachusetts Press
Amherst & Boston

Library of Congress Cataloging-in-Publication Data
Murphy, Priscilla Coit.
    What a book can do : the publication and reception of Silent spring /
Priscilla Coit Murphy.
        p. cm.—(Studies in print culture and the history of the book)
Includes bibliographical references and index.
ISBN 1-55849-476-6 (cloth : alk. paper)
1. Carson, Rachel, 1907–1964. Silent spring. 2. Authors and
publishers—United States—History. 3. Mass media in literature—United
States—History. I. Title. II. Series.
    QH545. P4M87 2005
    333.72'0973—dc22
                    2004019704

British Library Cataloguing in Publication data are available.

For Tom and Nick

CONTENTS

List of Illustrations   xi

Preface and Acknowledgments   xiii

1. *Silent Spring* and Its Contexts: "The Right to Know"   1

2. Author and Agent: "Where an Author Can Call His Soul His Own"   19

3. Editors and Publishers: Dealing with a "Super-Ruckus"   53

4. Opposition: "How Do You Fight a Best-Seller?"   89

5. Media: "One Formidable Indictment"   119

6. Audience: "This Ought to Be a Book"   159

Conclusion: "Speaking Truth to Power"   183

Appendix: Perspectives on the Study of *Silent Spring*   199

Notes   223

Index   251

ILLUSTRATIONS

FRONTISPIECE
Cartoon by Gordon Brooks from May 1963 *Yankee Magazine*

Rachel Carson, portrait   21

Rachel Carson at work   22

Carson with CBS's Eric Sevareid in her Silver Spring home   42

Paul Brooks, editor in chief, Houghton Mifflin   60

Advertisement for *Silent Spring*   78

Post-PSAC report advertisement for *Silent Spring*   82

Sample page from "List of Principal Sources," *Silent Spring*   86

Photo of book club from Oakmont, Pennsylvania, newspaper   123

Article from suburban New York newspaper group   125

Cartoon from *Des Moines Register*, July 26, 1962   132

Photo from *Life* magazine feature, October 12, 1962   142

TV Guide listing, March 30, 1963   151

Newspaper advertisements for *CBS Reports*   152

Editorial, Lynchburg, Virginia, *News*   157

Robert Darnton's Communications Circuit   201

PREFACE AND ACKNOWLEDGMENTS

"I have just heard that one of the chemical companies is now inserting its propaganda in the supermarket publications. A future social historian will be writing his Ph.D. thesis on the career of SILENT SPRING—just you wait!"

Houghton Mifflin Editor in Chief Paul Brooks,
writing to Rachel Carson, November 1962

When I first began to think about *Silent Spring*, I was simply looking for a news-making book that might be a good laboratory demonstration of how books and the media interact. From my list of about ten candidates, *Silent Spring* stood out mainly because it had been published both in magazine format and as a book, which I thought might offer a useful comparison of the functions of the two media. I was only distantly aware of the controversy *Silent Spring* had prompted, for it was published in 1962, while I was still in high school. Rachel Carson's name was known to me first because of her popularity with New England sea lovers of my parents' generation and later because she was named among the feminist pantheon of strong women. Other than that, I knew almost nothing about Carson herself, the drama of her life, or the vigor of the opposition to *Silent Spring*; and until 1996 I had never read *Silent Spring*.

The experience of reading *Silent Spring* for the first time was itself startling, even more than thirty years after the fact. Despite my scholar's intention of remaining as impartial on the issue of pesticides as possible, the work impressed me for its clarity, its passion, its lyricism, its logic, and its diligent scholarship—much as it had impressed readers in the 1960s. That it continues to impress others the same way is obvious from the fact that it has never gone out of print, as well as from the respect of the various notables who have added their own prefaces and commentary to later editions.

Nor has *Silent Spring* lost its ability to provoke controversy. There are some who still rail at the names of Rachel Carson and *Silent Spring*, some who still feel compelled to criticize her science and philosophy long after the fact, and some who classify her with that "destructive" group of sixties-era social critics whom they hold responsible for the deterioration of modern culture. Moreover, one of the chemical companies was still so tender on the subject of *Silent Spring* that when I wrote for permission to include a reprint of a piece from a house publication, a company lawyer called me early one morning to suggest I consider the possible harm to the company, its employees, and myself if I were to use the reprint. Never mind the fact that in 1962 thousands of copies of that very piece had been sent around the nation by the chemical companies themselves to newspaper and magazine editors, legislators and government officers, television producers, and influential citizens, or that the same piece had been reproduced in its entirety in several of the newspapers. What the companies were happy to have a matter of public record then embarrasses them now.

The first irony of that exchange was that the point of my study of *Silent Spring* has almost nothing to do with determining whether Carson and *Silent Spring* were right or wrong about pesticides; and I am the first to admit that I am no expert on the subject, although I have learned a certain amount from material written by those who are. What is more ironic is that the chemical company's response had the same negative effect on me in 1999 that the heavy-handed opponents of *Silent Spring* seemed to have had on much of the media and the audience in 1962. In many ways, such tactics backfired, making it difficult not to see the entire debate in terms of bad guys versus good guys (or women), the single citizen versus a powerful industry.

If this story—if this book about a book—seems therefore to lean a bit toward Carson's camp, I concede that some partiality evolved just because of what I learned along the way about the actions of Carson's rather frantic opposition. Remember, however, that this story is about a book and its author; and therefore it is probably inevitable that a history of the adventures and misadventures of the two should treat them as combined protagonist—the hero/heroine, so to speak. Moreover, this "book about a book" is first and last a history. I intended, of course, that the history might have some theoretical implications or at least might offer a case study against which theory may be tested; and so for those interested, the theoretical and methodological perspectives that informed my work are discussed in an appendix. The "public career" of *Silent Spring*, however, is a pretty good story all by itself, and it is as good a place as any to start thinking about the future of books, or at least such books-with-a-mission.

Before mentioning the many people to whom I am much indebted, I wish first to express a sense of deep, albeit troubled, gratitude for the existence of the archived material on Rachel Carson, Houghton Mifflin, and the *New Yorker*, and to those whose archiving impulses made that material available. Book history and media history in general are both much hobbled by the failure of organizations to preserve their documents and make them available to scholars. Much is being lost to posterity because no one wants to devote storage space or cataloging time, or because worries about proprietary interest persist long after its practical lifespan, or because the apparent efficiency of electronic media has replaced paper trails with ethers. I was fortunate to be studying an era prior to the technological transitions of the latter twentieth century, but had I chosen a book published only twenty years later, I would not have been nearly so blessed with primary material. Worse than ironic is the fact that the very organs by which we record our present for the future—newspapers, magazines, television, radio, film, and books—are being stripped of their own history by their producers' restriction and destruction of records.

Fortunately for this book, Rachel Carson's agent, Marie Rodell, was also Carson's ardent archivist, preserving and organizing an extensive collection of papers, manuscripts, and clippings (now located at Yale's excellent Beinecke Library), which were augmented by the *New Yorker* archives at the New York Public Library and the Houghton Mifflin archives at Harvard's Houghton Library. Therefore I first must thank Frances Collin for her permission to use the Rachel Carson Papers at Yale, without which this study could not possibly have taken place. As to the many librarians and their staffs in whose care the various archives reside, all scholars must be grateful for their existence and work. With much gratitude, I particularly wish to acknowledge the help first of Stephen Jones and all of the staff at Beinecke Library at Yale, where I spent many weeks in Rachel Carson's archived world. Similarly, I wish to thank those at the Houghton Library at Harvard, especially Elizabeth Falsey; the staff of the Manuscripts and Archives Division at the New York Public Library; the Davis Library Reference and Inter-Library Loan staff at the University of North Carolina at Chapel Hill; Joseph Schwarz at the National Archives and Records Administration; and those at the Hagley Library in Wilmington, Delaware.

I wish I could say thank you in person to the late Paul Brooks, Rachel Carson's friend and editor at Houghton Mifflin, who passed away shortly after our last conversation, for his insights, memories, and friendly demeanor toward yet another *Silent Spring* scholar. I am grateful to his family for allowing access to him in his latter months and for their efforts to find me a photograph of him to share with my readers.

To the late Margaret Blanchard of the University of North Carolina at Chapel Hill, I am much indebted for her First Amendment perspectives on my eccentric interest in book publishing as part of media history, as well as for her patience, flexibility, good counsel, and eagle-eyed reviews of early versions of the manuscript.

In addition, I thank the American Antiquarian Society for including me, with support, in a summer institute that introduced me to the field of the history of the book. I am also grateful to the Association for Library and Information Science Education for its support.

Others to whom I owe thanks include Diana Post at the Rachel Carson Council for her assistance with some critical material and for showing me Rachel Carson's Silver Spring house; Ann Claybrooke for saving me from microfilm vertigo; Deborah Applegate for logistical support at Yale; and Melissa Newt for her eleventh-hour help.

I am most grateful to the University of Massachusetts Press for their support and effort on behalf of this book, especially Carol Betsch for her expert and interested shepherding through the publication process, and to Karen Bosc for her careful and respectful editing of the text.

I am, above all, especially happy to acknowledge my great debt to Paul Wright of the Press for his belief and interest in the project from the start, his professional counsel to me at critical junctures, and his friendly advice along the way.

Finally, I owe something much beyond thanks to my patient and encouraging family, my heroes, Tom and Nick Murphy.

What a Book Can Do

CHAPTER 1

Silent Spring and Its Contexts: "The Right to Know"

This is the story of a news-making best-seller about the unlikely topic of chemical pesticides. Rachel Carson's 1962 work, *Silent Spring*, prompted a vigorous public controversy, both as a prominent Houghton Mifflin book and as a *New Yorker* serialization that actually preceded the book by three months. *Silent Spring* described the toxic side effects of commonly used insecticides, herbicides, and fungicides; and it criticized what Carson felt were indiscriminate, irresponsible, and finally dangerous practices and policies. Government agencies, the chemical industry, agribusiness, and even academic researchers were the targets of her criticism. In response, an organized opposition mounted a countercampaign, in place even before publication, to discredit the work and to limit its impact on the public. Those tactics included pressure to stop publication and to control media coverage.

For more than a year, the controversy raged publicly throughout the media, in editorials, opinion columns, news articles and broadcasts, book reviews, and letters to the editor. From garden club meetings to the floor of Congress, *Silent Spring* drew its advocates and its detractors into sometimes acrimonious debate, one not only covered by the media but engaging the media as participants as well. Remarkable as it is that one author was able to set forth her message and to reach audiences at every level—from neighborhood to statehouse to White House—it is even more remarkable that she was able to do so through a medium whose paradigm is the simple dyad of author and single reader.

At the turn of the millennium, many prophets were predicting the end of "the book as we know it," even as many found themselves trying to arrive at a precise definition of "the book as we know it." But defining what a book *is* does not give us the information needed to consider whether books will actually survive. The physical entity "book" can be defined and redefined to argue the issue either way—that bound, printed books of a certain size and shape are already outmoded, or that books will always exist in some form, just not necessarily as paper and ink.

What a book *does*, however, can tell us what we use books for, giving us the opportunity to consider, among other things, whether any other medium would perform the same function or fill the same needs better than, or even as well as, a book. On occasion, the impact of individual books such as *Mein Kampf* or *Uncle Tom's Cabin* have been woven into social, cultural, or political history; but the approach taken has commonly focused on the individual author as the pivotal, history-making actor. That approach has glossed over the dynamics by which that author's message becomes public and the public receives and reacts to that message. In the course of the twentieth century, that process unavoidably and increasingly involved the elaborate and far-reaching mass media system that now dominates American social and political culture. Thus examining what a book does means inquiring about how the book and the media system interrelate and interact: What distinguishes a book from other media? What happens to books when they emerge into the media system? And what happens in the media system when a news-making book appears?

In an era when the presence of a mass media system has long since become a defining characteristic of American politics and society, the ability of one citizen to have his or her voice heard by the masses has seemingly disappeared without access to high-priced, highly managed, national media institutions. Yet as Philip Meyer suggested, "The book author has become the modern equivalent of the lonely pamphleteer."[1] Authors in every field with every purpose, even journalists and public figures who already have direct access to the media, have continued to resort to books to communicate their messages to the public at large. Moreover, book publishing, far from collapsing with the advent of each new "competitor" medium, has continued to survive and even to flourish. Why should this be?

Here, we will review and analyze the history of the single work *Silent Spring* and hope along the way to uncover some different kinds of answers to that question. This is not an examination of the environmental, literary, or rhetorical issues surrounding the book, though all are relevant and will come into play. This is a study of the genesis and conduct of the public debate around *Silent Spring*, from Rachel Carson's earliest expectations to the climactic *CBS Reports* program, "The Silent Spring of Rachel Carson." Although some of the story necessarily takes place behind the scenes, it is the public life—or what Houghton Mifflin editor Paul Brooks called the "career"—of *Silent Spring* that will tell us the most about how books may function in our elaborately mediated culture. Although the events took place over four decades ago, many of the mechanisms, relationships, and perceptions about authors, books, and the media persist; and in fact, our ability to assess the future status of books may hinge on identifying exactly where change has taken place and where it has not. But first we must dissect those mechanisms, relationships, and perceptions.

In the coming chapters, we will look at what Rachel Carson believed she was doing when she wrote and published *Silent Spring*—how she felt about books, the media, and her audience. We will look at those who worked with her: her agent, her publishers, and her editors. We will look at what upset her opponents and how they chose to meet the challenge of her book, as well as how both sides approached the media. We will explore how the media themselves saw their role in the debate. And finally we will look at some of the perceptions and responses of the public. Before undertaking that long circuit, however, it is important to know about the circumstances and actual content of *Silent Spring*, whose official publication date was September 27, 1962.

The Making of *Silent Spring*

Rachel Carson had been interested in pesticides, DDT especially, as early as 1945; but a 1957 lawsuit concerning spraying over Long Island to eradicate gypsy moths prompted her to consider seriously the timely need for a magazine article on pesticide hazards. She wrote to fellow *New Yorker* writer E. B. White, urging him to report on the court case, but he declined. Deciding to take on the issue herself, Carson began discussions with both the *New Yorker* and Boston's Houghton Mifflin, the publisher for her most recent book, *The Edge of the Sea*. In May 1958 she signed virtually simultaneous contracts with both the magazine and Houghton Mifflin. The original plan entailed having Carson act as editor for a book on the effects and hazards of chemical pesticides, with pieces written by several authors and at least one chapter by Carson herself that would also constitute a *New Yorker* article. She had already accumulated limited material on the subject; but as her research began in earnest, she quickly determined that the entire book could be written only by one author—herself.

Through the ensuing three years, Carson suffered frequent debilitating illness and family misfortune. At the same time, her research was mushrooming in scope and depth. Thus, writing and publication were delayed considerably. As 1961 came to a close, Carson was frantically trying to finish writing and revising her manuscripts for both the *New Yorker* and Houghton Mifflin. The piece for the *New Yorker* was eventually split into three segments; and publication was set for the June 16, June 23, and June 30 issues, three months prior to Houghton Mifflin's official book publication date. Very shortly after the first *New Yorker* segment appeared, the media took notice, as did the chemical companies. While editorials and letters began to appear in newspapers, for the most part acclaiming the article, the agricultural chemical industry mobilized a substantial public-relations effort with a $250,000 war chest to counter the articles' message. Midway through publication of the series, a lawyer contacted the *New Yorker* to admonish the magazine, and in August the same lawyer, now identifying himself as representative of chemical producer Velsicol, contacted Houghton Mifflin in an unsuccessful attempt to stop publication altogether. Not long after, during a presidential press conference, a reporter asked President John F. Kennedy if he had considered asking the U.S. Department of Agriculture (USDA) or the Public Health Service (PHS) to look into pesticide dangers. His affirmative response specifically mentioned "Miss Carson's book," even though the book itself had yet to be published. At Kennedy's behest, the Life Sciences Panel of the President's Science Advisory Committee (PSAC) directed its attention to the issue of pesticide use and abuse. By mid-September of 1962, still well before the official publication date, the book had received considerable media exposure beyond the usual preliminary publicity in book reviews and announcements.

Meanwhile, Houghton Mifflin was already enjoying considerable rights success with *Silent Spring*, which had been announced as the Book-of-the-Month Club's October selection. Consumers Union sought and received permission to put out a special edition for its members; and numerous periodicals sought serialization or reprint rights. Noting the growing media and industry interest, Paul Brooks eventually ordered an exceptionally large original print run of 100,000 copies, and by December 1962, 500,000 copies were in print in various forms. Retail sales were at least as gratifying as rights sales. Two weeks after the publication date, *Silent Spring* first appeared on best-seller lists; and by the first week in November it had reached first place, which it held until John Steinbeck's *Travels with Charley* moved it to second place in mid-December.[2] It numbered among the top ten best-sellers for six months, showing brief resurgence in April and May of 1963, when two events renewed attention to *Silent Spring*: a television broadcast and the report of the presidential science committee.[3]

As *Silent Spring* was filling bookstore shelves, CBS News was negotiating with Carson's agent, Marie Rodell, for Carson's appearance on the prestigious *CBS Reports* program. Originally planned for November, the broadcast was tabled until spring of 1963 for reasons not entirely clear. The program aired on April 3, despite withdrawal of advertisements by some sponsors and threats by others to do the same, as well as a warning from the same chemical industry lawyer who had contacted the *New Yorker* and Houghton Mifflin on Velsicol's behalf. A special follow-up segment was aired in mid-May on the release of the PSAC report, widely interpreted as vindicating *Silent Spring*.

Unfortunately, Rachel Carson's health during the flurry of success and publicity was deteriorating. She had known since 1960 that she had breast cancer and that her time was short, but she had kept that information from nearly everyone. She severely limited personal appearances but appeared at two congressional hearings on pesticides in June 1963. The Fawcett paperback edition of *Silent Spring* appeared in early 1964, and another flurry of attention followed, sadly punctuated in April by Carson's death a month short of her fifty-seventh birthday.

This brief outline of the events surrounding *Silent Spring* will be filled in during the chapters to follow. To understand exactly what so concerned the public and so upset the chemical industry, agribusiness, and government agencies, we obviously need to know what the book said. However, as it turned out, we must also know what the book did *not* say but was imputed by its opponents to have said—as well as what the difference was between the *New Yorker* articles and the book.

What the Book Said and What It Did Not Say

Silent Spring begins with a "Fable for Tomorrow" (which would be parodied by several critics) describing a town "in the heart of America" that had suffered "a strange blight" that killed animals and sickened or even killed families with "mysterious maladies." The result is "a strange stillness.

The birds, for example—where had they gone? . . . It was a spring without voices." The only clue was a "white, granular powder" that some weeks earlier had "fallen like snow upon the roofs and the lawns, the fields and streams." In an indirect comparison to radioactive fallout, Carson explains that "no witchcraft, no enemy action had silenced the rebirth of new life in this stricken world. The people had done it themselves." She concludes the chapter saying that the town does not yet exist but that it easily could, and indeed several communities had already suffered one or another form of the disaster. She writes, "What has already silenced the voices of spring in countless towns in America? This book is an attempt to explain."[4]

Her second chapter—which would be pivotal in the controversy—defines the problem as Carson saw it, presents her philosophical reasons for writing the book, and makes explicit that she does not argue for a complete ban of pesticides. "We have subjected enormous numbers of people to contact with these poisons, without their consent and often without their knowledge. If the Bill of Rights contains no guarantee that a citizen shall be secure against lethal poisons distributed either by private individuals or by public officials, it is surely only because our forefathers . . . could conceive of no such problem." At the chapter's end, she quotes Rostand: "The obligation to endure gives us the right to know."[5]

The next chapters pursue the proposition that chemical pesticides—agents to control insects, plant diseases, and weeds—present a threat that had been inadequately investigated, for both societal and economic reasons.

> There is still a very limited awareness of the nature of the threat. This is an era of specialists, each of whom sees his own problem and is unaware of or intolerant of the larger frame into which it fits. It is also an era dominated by industry, in which the right to make a dollar at whatever cost is seldom challenged. When the public protests, confronted with some obvious evidence of damaging results of pesticide applications, it is fed little tranquilizing pills of half truth. We urgently need an end to these false assurances, to the sugar coating of unpalatable facts. It is the public that is being asked to assume the risks that the insect controllers calculate.[6]

Pointing out the commonalities of effect and consequence between insecticides and herbicides, she expands the discussion to pesticides in general, making the point that the term "pest" is relative—what is a pestilence in one situation may be a critical need in another.

Her third chapter, "Elixirs of Death," discusses the chemical formulas and properties of the various insecticides, explaining how such chemicals as DDT harm all living organisms in the same way that they harm or kill their intended victims. "Surface Waters and Underground Seas" next describes the continuity of earth's aquatic system of rivers and oceans and how pollution is thereby a potentially uncontainable evil. "It is not possible to add pesticides to water anywhere without threatening the purity of water everywhere."[7] The fifth chapter, "Realms of the Soil," lays out the genesis and life of the soil, dependent as it is on organisms living as well as dead. "Earth's Green Mantle" refers to the "web of life in which there are intimate and essential relations between plants and the earth, between plants and other plants, between plants and animals."[8] In "Needless Havoc," Carson turns to the frequency with which "eradication" campaigns against, for example, the Japanese beetle had been completely unsuccessful after all—causing ecological mayhem but failing to get rid of the targeted problem and even creating in some cases particularly favorable conditions for increasing abundance of the targeted "pest."

The eighth chapter, "And No Birds Sing," was the inspiration to her editor and publisher, Paul Brooks, for the title of the book. It begins, "Over increasingly large areas of the United States, spring now comes unheralded by the return of the birds, and the early mornings are strangely silent where once they were filled with the beauty of bird song."[9] The next few chapters follow essentially the same format—presenting one aspect of the problem, providing explanations and illustrative incidents, and concluding with exhortations to acknowledge the problem and demand solutions. "Rivers of Death," which focuses on perils and catastrophes among fish life, concludes that the threat to fisheries "by the chemicals entering our waters can no longer be doubted." Carson asks, "When will the public become sufficiently aware of the facts to demand . . . action?"[10]

The next chapter, "Beyond the Dreams of the Borgias," carries some of Carson's most confrontational material, and the title's image would recur in media coverage. The chapter addresses the hazards of aerial spraying against gypsy moths and Japanese beetles as well as crop dusting. In it, Carson challenges the Food and Drug Administration (FDA) on the issue of contamination of consumer foodstuffs, comparing the situation of the public to guests of the Borgias, having no idea what sort of poison might be present in their food. "To the question 'But doesn't the government protect us from such things?' the answer is, 'Only to a limited extent.' "[11]

The next three, somewhat more technical, chapters detail the physiologic and cellular effects, often delayed, of ingestion of pesticides into the human body. "One in Every Four" addresses cancer specifically (noting, among others, studies dealing with the consequences of arsenic used by tobacco growers). "Nature Fights Back" describes processes by which nature adapts to and overcomes threats to survival, including mutations and increased resistance on the part of the "pests" targeted by chemical pesticides. The sixteenth chapter suggests the possibility that resistant strains of insects might eventually overwhelm known methods of control and that epidemics of once-controlled diseases such as malaria could recur with devastating effect. In her concluding chapter, she discusses the scientific possibility of alternative, nonchemical, natural means of pest control.

Throughout her discussion, Carson frequently raised questions about why pesticide abuse had not yet been well investigated or discussed, why research into less toxic means of pest control had been limited or ignored, and who was responsible for the failure to protect the general public. Her view of the relationships among industry, government, and research was clear, as was her attitude regarding the economic priorities of the era. Concerning the "otherwise mystifying fact that certain outstanding entomologists are among the leading advocates of chemical control," she explained: "Inquiry into the background of some of these men reveals that their entire research program is supported by the chemical industry. Their professional prestige, sometimes their very jobs depend on the perpetuation of chemical methods. Can we then expect them to bite the hand that literally feeds them?"[12]

Silent Spring is, overall, a call to action based on a carefully delineated explanation of the threat—current and future—of damage to life by misuse of pesticides. "The public must decide whether it wishes to continue on the present road, and it can do so only when in full possession of the facts."[13] The underlying logic of her argument follows from the concept of natural interrelatedness of all living things and the need to sustain those relationships in a dynamic balance. The shorthand for that idea—"the balance of nature"—was a term used before Rachel Carson's era; but until *Silent Spring*, it carried connotations of irrational sentimentality and was applied derisively to the beliefs of conservationists and nature lovers. The legacy of *Silent Spring* was not only to have prompted debate and action on the specific issue of pesticide abuse but also to have made vivid, accessible, and acceptable the idea that nature requires balance, an idea that formed the basis of popularized environmentalism in the coming decades.

Finally, two characteristics of *Silent Spring* must be kept in mind as its career in the public eye is explored. First are the appended fifty-five pages of bibliography and source material, unprecedented in any of Carson's previous works or in any other popular nature books of the time. Prominent in the front matter is an "Author's Note": "I have not wished to burden the text with footnotes but I realize that many of my readers will wish to pursue some of the subjects discussed. I have therefore included a list of my principal sources of information, arranged by chapter and page, in an appendix which will be found at the back of the book."[14] The presence of that reference material was to play a pivotal role in criticism and defense of the book. With respect to the conduct of the debate, even more important is the fact that Carson explicitly declined to prescribe total abstinence from pesticides. Early on, in the second chapter, she makes the following statement (emphasis added): "*It is not my contention that chemical insecticides must never be used.* I do contend that we have put poisonous and biologically potent chemicals indiscriminately into the hands of persons largely or wholly ignorant of their potentials for harm."[15] Many of her critics chose to ignore that disclaimer, however, as well as the proposed alternatives discussed in her final chapter.

Book versus Magazine

At first glance, the *New Yorker* articles appear to present essentially the same material as the book, opening with the parable of the unnamed, birdless town and progressing through the same points covered in the book chapters but lacking their headings. The phrasing and terminology seem unchanged, certainly not "dumbed down" in any way; on the contrary, some quite technical discussions of biochemical information are included, particularly in the second installment. But closer inspection reveals some telling differences in organization.

The first of the three installments focuses on Carson's opening parable and the ravages of pesticide abuse, using anecdote and Carson's uniquely forceful language—overall giving a rather dramatic first impression. The second installment moves into the more technical explanations and a discussion of natural interrelationships. The third offered evaluations and arguments in a somewhat more philosophical vein. Not until the end of third installment, however, does the reader come upon the all-important disclaimer that Carson does not argue for complete abstinence from pesticide use. Moreover, her examination of alternative methods of pest control, discussed at some length in her final chapter, scarcely appears in the magazine version of *Silent Spring*. Finally, because footnotes and bibliography were not appropriate to the magazine format, references to the scientific work on which Carson based her arguments were worked into the text in the form of researchers' names or institutional locations. The detailed specifics of supporting documentation are absent, as are the fifty-five pages of references.

Overall, the effect and the rhythm of the pieces were—appropriately—that of a investigative, journalistic feature, beginning with dramatic themes and anecdotes, supported in the middle mainly through the accuracy and lucidity of Carson's explanation of scientific phenomena, and concluding with philosophical commentary. Those who reacted only to the first installment of *Silent Spring* were reacting to a message that could have been interpreted as advocating a complete ban and, furthermore, a message with uncertain scientific basis. Those who read the remaining two installments were still without the full documentation present in the book, nor were they given a full picture of potentially viable alternatives to chemical pest control.

Once the book came out, of course, reviewers, editors, and readers had access to the full discussion, complete with disclaimer, sources, and proposed alternatives. The public argument had already begun, however; and with the notable exception of a few members of her opposition, no one made a point of the differences between the two media. Rather, it was the book *Silent Spring* that was considered the document of record and the primary carrier of Carson's message into public awareness. To better understand the public, corporate, and governmental responses to that book, it is worth reviewing briefly the historical and cultural contexts in which it emerged.

America in the 1960s

When *Silent Spring* appeared in 1962, it was greeted by an American population enjoying unprecedented wealth, literacy, and general education. At the same time, it was a nation released from intensely immediate concerns about wartime survival and now able to turn attention to the conduct of its own public life, including the ramifications of the ongoing scientific revolution for humanity and nature. Cold War anxiety had brought with it worries about environmental threats from radioactive fallout from atomic bomb testing, and many in the *Silent Spring* debate, including Carson herself, drew explicit parallels between the dangers of fallout and the hazards of pesticidal chemicals. Strontium-90 had been found in the milk supply; and the food industry had been shocked by the "cranberry scare" of 1959.[16] Less than three weeks before Thanksgiving, Secretary of Health, Education, and Welfare Arthur S. Sherwood had called a press conference concerning reported contamination of northwestern cranberries by aminotriazole, a weed killer reported to cause cancer in rats. He advised "housewives" not to use cranberries until and unless tainted lots could be distinguished from the untainted. The cranberry industry endured a catastrophic collapse in national sales unmitigated by belated arrangements to certify safe lots of cranberries. The memory of that scare would be held in the public mind as a message that pesticides could harm humans, that they could find their way into the food supply, and that the government had a mandate to protect consumers. Perhaps even more forcefully, the episode would be held in the memory of agribusiness as an object lesson about the devastating cost of public panic in response to such public health alerts.

Not irrelevantly, farmers were at the same time confronting a set of circumstances that the media dubbed the "farm crisis." Those who should have been beneficiaries of the postwar population explosion were pressed by extremes of protracted drought and unusual floods, along with changes in farming technologies that were putting the small family farm in peril. The rise of agribusiness reflected structural changes similar to those in the nation's other industries, as large overtook small and technology encroached on labor. Expensive, heavy farm machinery had become necessary to meet the increasing demands of a growing and wealthy nation; and productivity had to be increased through a vigorous growth in agricultural science—notably the uses of chemical fertilizers, herbicides, and insecticides. Farmers depended more and more on these chemicals, but the farmers were also the ones most immediately vulnerable to any deleterious effects. The media covered in somewhat sentimental fashion the impending demise of the family farm, but it was not until *Silent Spring* appeared that reports of pesticide-related health problems among farmers were publicized beyond their local press.

Meanwhile, through widening use of media overall and television in particular, the general public was becoming more aware and more involved in social issues; and the media themselves were turning attention to public affairs in new ways. Although television news was still somewhat in its infancy, often more local than national, the era marked the heyday of the television documentary. In his classic 1961 speech deploring the "vast wasteland" of television programming in general, FCC chairman Newton Minnow made explicit exception of television documentaries and information programs,[17] singling out *CBS Reports*, which had become a gold standard for the industry with such exposés as "Harvest of Shame," and which would soon turn its brilliant spotlight on *Silent Spring*.

Despite the rise of television, reading—of books as well as newspapers and magazines—was still a primary means of receiving communication. The numbers and types of magazines had begun a remarkable expansion, particularly in domestic niches such as house and garden magazines. Although daily newspapers were declining in the face of broadcast journalism's ability to provide more immediate and more vivid news, the number and circulation of Sunday newspapers actually rose, particularly in middle-sized cities with rapidly spreading suburbs.[18] To these markets, the Sunday papers delivered a wealth of opinion columns, book reviews, gardening pages, personal portraits, and letters from readers—all very much involved in the *Silent Spring* discussions. For its part, book publishing was still enjoying the postwar boom, particularly in nonfiction; and established houses like Houghton Mifflin had not yet been profoundly affected by the changes in management practices and market principles that would soon demand consolidation, conglomeration, merger, acquisition, and the requirement that each book be a blockbuster.

Whether through education, print or broadcast media, the amount of information available to the American public was increasing exponentially. And this was a public disposed to be informed and involved, as parents, consumers, and voters. Books like David Riesman's *Lonely Crowd* (1950), C. Wright Mills's *Power Elite* (1956), or Vance Packard's *Hidden Persuaders* (1957), *Status Seekers* (1959), and *Waste-Makers* (1960) called attention to materialism and other troubling aspects of contemporary American society, and the public's interest in them evidenced a new openness to social criticism, as well as a creeping distrust of business and government.

Arguably, that openness was the foundation on which the political and social activism of the sixties would rest. Despite the frequent characterization of the fifties as smugly conservative and conformist, public-minded action was still part of the American value system, perhaps residual from the war effort, perhaps of more essential origin as part of national self-definition. Regarding the atomic bomb in particular, Paul Boyer noted, "With a high level of anxiety feeding into a diffuse but strong impulse to action, earnest advocates of a bewildering variety of causes, projects, and points of view now began to compete for the public ear."[19] The implication for the public was that citizens could, and should, take action that would sway policymakers. The increased sense of global citizenship following the global war of the 1940s had, moreover, created a sense of peculiarly American responsibility to take moral and scientific leadership. In discussing the movement among scientists against further development of nuclear weapons, Boyer quoted Albert Einstein as saying, "To the village square, we must carry the facts of atomic energy. From there must come America's voice."[20]

In the national political arena, a certain amount of heightened rhetoric might have been expected in any midterm election year, as 1962 was. But grassroots activism was burgeoning, notably in the form of the civil rights movement; and a sense of the general force of public conscience and civic action was thus reviving—if it had ever in fact slept. Some might argue convincingly that anticommunist activities, even taken to McCarthyite excess, had been prompted by the same impulse. That impulse was rooted in the belief that the American public should be informed and alerted— that the public had a fundamental "right to know"—and further, that on the basis of certain kinds of information, the public could and should be mobilized to act to improve the system. In that framework, Rachel Carson joined a long tradition of public-spirited sentinels like Harriet Beecher Stowe and Sinclair Lewis. Her message was presented by a woman already known for her expertise and love of nature, to an audience able to receive information both technical and critical of policy and practice, in an increasingly complex media system capable of reaching the broadest and the most local audiences.

Moreover, that message was greeted by a population already acquainted with the idea of dangers in the environment, even if the specific threat posed by pesticides was news. Environmentalism itself had indeed existed well before Rachel Carson's emergence as its standard-bearer, in the form of conservation groups and naturalist interests that had been in place since the turn of the century or even earlier. In the postwar era of peace and plenty, garden clubs, nature groups, and conservation organizations grew both in number and activity. According to stereotypes of the day, leisure time involved sending the men out hunting and fishing in field and stream, while the women tended the gardens around the house, now affordable thanks to the GI Bill. Whether or not the stereotypes held true, outdoor activity was on the rise, and the suburbanization of the American population meant a boom in ownership of homes with yards to go with them. But at the same time, the automobile and the industrial boom were creating increasingly evident environmental problems. The words "ecology" and "environmentalism" were not yet in common use ("conservation" was the more common term); but the term "pollution" with reference to air and water was gaining popular currency as John F. Kennedy took office, and in fact Dwight Eisenhower had already initiated study aimed at reducing water pollution. News stories on smog and polluted rivers and lakes were becoming more frequent, though not yet with the fearful urgency of discussions of radiation and fallout; and chemical contamination from pesticides had already been noted in several quarters well before *Silent Spring* appeared.[21]

Environmental historians commonly note that fact, and they characterize Carson's book as a watershed event that catapulted the issue into the view of a previously ignorant general public, thereby popularizing environmentalism. Historically, the before-and-after picture is not quite so clearcut, even though Carson herself described her role as enlightening a citizenry kept purposely in the dark by industry and government. Most histories of the pesticide issue[22] trace early concerns from the mid-1940s, immediately following the widespread use of DDT during the war, when Carson herself attempted unsuccessfully to interest the *Reader's Digest* in an article on DDT's hazards. But other publications were raising questions, and an opinion piece in a 1946 *New Republic* opened its discussion in a style and language remarkably similar to Carson's a decade and a half later:

> On May 23, 1945, the sun shone warmly on a large oak forest near the village of Moscow, Pennsylvania. Bird calls and songs rang through the woodland as the birds flew about feeding hungry young ones. But the forest was ill; its leaves were covered with millions of devouring gypsy-moth caterpillars. . . . Early the next morning, an airplane droned over the forest, dropping a fine spray of DDT in an oil solution at the rate of five pounds an acre. The effect was instantaneous. The destructive caterpillars, caught in the deadly rain, died by the thousands. On May 25, the sun arose on a forest of great silence—the silence of total death. Not a bird call broke the ominous quiet.[23]

The silenced-birdsong image was a common theme in the literature challenging wholesale application of chemical pesticides. Sometimes the image was an allusion to the Keats poem Carson herself used ("The sedge is wither'd from the lake/And no birds sing" from *La Belle Dame Sans Merci*), but sometimes it described personal experience with the aftereffects of spraying.

Before the early sixties debate around *Silent Spring*, certain events had drawn the attention of the national and local media to the issue of pesticide abuse. In the spring of 1949 a DDT "scare" rose out of veterinary studies of ill effects in animals. That controversy took very much the same form as the one to come in 1962, with DDT producers and government officials charging "hysteria" and declaring DDT not only safe when properly used but also essential to an adequate food supply. By the late 1950s questions were being raised widely by scientists, naturalists, and conservationists, notably the National Audubon Society, whose popular *Audubon Magazine* carried many and frequent pieces about the harmful effects of chemical pesticides. A pivotal event, for Carson in particular but also for many others concerned about pesticide use, was the 1957 lawsuit brought by Long Island residents against the U.S. Department of Agriculture (USDA) for its aerial spraying over their communities to kill gypsy moths. Led by ornithologist Robert Cushman Murphy, the group lost the suit, and the Supreme Court denied review on technical grounds: spraying had already taken place.[24] The proximity to New York City meant the story could easily be inserted into the general media stream, and Justice William O. Douglas's dissent in the Long Island case was reprinted in the *Saturday Review.*[25] Not long after came the 1959 cranberry scare.

Local press attention around the country took note of increased spraying programs because of circumstances as diverse as Dutch elm disease, an encephalitis epidemic, highway safety and beautification, and postwar demands for increased food production. Those in the areas sprayed sometimes raised public objections to the aftereffects, particularly in the Midwest. In the spring of 1962, Chicago and Cleveland were confronting Dutch elm disease while the *Des Moines Register* ran articles and letters about Iowa roadsides denuded of foliage and wildlife casualties as a result. Fears of summer epidemics of insect-borne diseases drove anti-mosquito spraying projects aimed at community lakes and ponds; the spraying sometimes prompted irate letters to the editor from neighbors.

Although local reports of pesticide problems otherwise appeared intermittently in farm area media, discussion was also ongoing in some eastern mainstream periodicals. The *New York Times* covered the issue on a continuing basis, beginning with its reports on early government inquiry into pesticide hazards substantially before *Silent Spring.* The *Times* coverage expanded in 1960 to include reports on pesticide pollution of water, as well as both reports and commentary on a federal bill to require coordination between the U.S. Fish and Wildlife Service (FWS) and government eradication programs. In July of 1961, a year before the *Silent Spring* series would appear in the *New Yorker,* a *Times* editorial called for controls and education regarding pesticide hazards.[26] Also in 1961, the *Saturday Evening Post* ran an editorial titled "Pesticides Are Good Friends, But Can Be

Dangerous Enemies if Used by Zealots," recounting Michigan professor George Wallace's study of the devastation to the campus's robin population and his assertion that "the current widespread and ever-expanding pesticide program poses the greatest threat that animal life in North America has ever faced."[27] In response to some of this outcry, the USDA felt the need to produce a public service film called *The Fire Ant on Trial* justifying and even glorifying its fire ant eradication project, a film that infuriated many conservationists, including Carson.

Nonetheless, although most Americans were probably aware of government pesticide programs and may well have had their own "bug bombs" for home use, the startled public response to the news about *Silent Spring* strongly suggests a general unawareness of possible hazards. Comparing the sheer volume of press attention to pesticides—pro, con, or neutral—before and after the appearance of *Silent Spring* reveals an unmistakably sharp increase immediately following its appearance. Some analysts have inferred that the preliminary media attention had served to "soften up" the public in a two-step process involving the elite press represented by the *New York Times*, such that *Silent Spring* functioned as accelerant rather than ignition in the DDT controversy.[28] An equally valid interpretation is that the media attention indicated a predisposition on the part of certain members of the media regarding *Silent Spring*, viewing it as additional evidence and ammunition to validate or extend what they already had decided was an important news issue. Although Rachel Carson was, thus, not the first to sound an alarm about chemical pesticides and spraying programs, nor even the first to try to write a book about it, her approach met a ready audience, one that included her publishers, the media, the general public, and even—though not happily—her opposition.

Approach and Issues

The fact that *Silent Spring* should appear in book form was significant to all who became participants in the debate it stimulated. Among those participants were not only author and reader but also many others in between: Carson's agent, editors, and publishers; her detractors, including pesticide producers and users, as well as government agencies charged with controlling pests and supervising pesticide use; the media that reviewed the book, reported on the controversy, and even took part in it; and the public consumers of those media, who may or may not have read *Silent Spring*—and who will be called its "audience" rather than its "readership" to reflect that fact. In this book I visit each of these participants in turn—from author to reader and those in between—to discern their perceptions about and approaches to *Silent Spring*. I ask two questions: What did it mean to them that *Silent Spring*'s message came in book form? What difference did that fact make in their plans, their responses, their actions?

One other group—state and federal legislators—were arguably the ultimate target audience for both sides of the debate. However, although the issue was taken up by Congress and many state legislatures by 1964, my focus here is on what happened that brought the debate to those halls. Indeed, behind much of this discussion is a conception of a public arena that draws on Jürgen Habermas's notion of a "public sphere" (notwithstanding Habermas's protestations that a public sphere ceased to exist with the advent of mediated public communication) in which a population sustains a collective consciousness of itself as a publicly acting, policy-determining entity.[29] The expectations not only of the reacting public, discussed in chapter 6, but also of the author herself with respect to the public have particular significance viewed in this light—the issue being the role of a book in the genesis and conduct of public debate on policy. (Further discussion of some philosophical and methodological underpinnings of my work is undertaken in the appendix.)

Finally, a note about the time frame of this analysis. Although Carson was interested in the harmful effects of DDT as early as 1945, her work on *Silent Spring* did not really begin until 1958. The articles came out four years later, in June of 1962, and the book three months thereafter, on September 27. Press attention came immediately with the appearance of the first *New Yorker* installment; it ballooned over the summer and kept up a steady pace for the next several months. A resurgence of interest came in mid-1963 with the landmark *CBS Reports* broadcast in April, the release of the PSAC report in May, and the beginning of congressional hearings in June. The paperback was released in early 1964. Carson died in April of 1964. My exploration here, therefore, begins more or less in 1958 and ends with Carson's death, with necessarily closer attention to events in the year between June 1962 and June 1963.

The reverberations of *Silent Spring* continue, however. The book is still in print, and a new paperback edition was published in 1994 with a foreword written by then vice president Al Gore. As my book was in preparation, controversies were roiling over anti-mosquito spraying to prevent the spread of the West Nile virus and equine encephalitis, as well as about genetically engineered grains damaging to wildlife but designed to be particularly responsive to specific producers' pesticides. Broader environmental issues such as global warming are ever more in the news, while Bill Moyers's televised *Frontline* study of chemical company activities surrounding the toxicity of polyvinyl chlorides suggests a persisting continuum of approach and behavior. The historical standing of *Silent Spring* as a pivotal moment in public discussion is still recognized by those as diverse as the *New York Times* in its list of the twentieth century's one hundred most important books, environmentalists like Gore and Moyers, and modern critics such as *Washington Times* editor Kenneth Smith, who castigated *Silent Spring* as the beginning of a troublesome and dangerous era in American democracy.[30]

Whatever one's attitude toward the content of Rachel Carson's book, the fact that it provoked a vigorous and abiding debate is incontestable. Almost as certain is that the debate was conducted within an assumption of free speech, protected by the First Amendment and extended to all media. With respect to the role of books within that media system, what is of interest here is not only what *Silent Spring* "did"—what it accomplished for its author, publishers, detractors, reviewers, discussants, and readers—but also what that accomplishment may say about our perceptions about books in our culture, a culture rife with alternatives to the reportedly moribund medium of books.

CHAPTER 2

Author and Agent: "Where an Author Can Call His Soul His Own"

Rachel Carson was, of course, already a writer of books. What she wanted to communicate in *Silent Spring* differed, however, from her earlier rhapsodic explorations of nature in that it was a warning, one aimed not just at a nature-loving readership but at the public in general. As an established author, she could have communicated via a magazine article, a newspaper feature or commentary, or perhaps even a scholarly monograph. She could even have acted outside of the media, as an activist with conservation groups or on her own; she was already well-connected. In point of fact, her choice was to write a book and magazine article simultaneously; but they had somewhat different functions for her, and she most often spoke of *Silent Spring* as a book.

Looking at how the author thought about her mission and how using a book fit into that mission not only opens the first act of the *Silent Spring* drama but begins discovery of some significant ways that books function in a media-complex society. Ultimately we shall ask what difference it made to Carson that her message would be published as a book—in her presentation of the message, in how she approached the public, and in how she dealt with those challenging her message. One or two rather obvious reasons for the almost automatic choice of the book form might immediately spring to mind, but the complex array of obvious and not so obvious factors is worth exploring.

The backdrop for the public drama that consumed Carson's last few years was the private drama in her life and persona, including her professional style and relationships. Among those relationships, that with her agent and friend, Marie Rodell, is of special significance. Rodell's role in bringing *Silent Spring* to light and in managing its public career was at least as pivotal as that of Houghton Mifflin's publicity department. Moreover, Rodell shared Carson's mission, which may readily be characterized as journalistic at core. Out of convictions adopted from Jean Rostand—"The obligation to endure gives us the right to know"—flowed Carson's research, her writing, and her consent to public exposure despite the many reasons for her deep aversion to it. Her conception of her mission, moreover, was based in—and the basis for—her relationship to her audience, the public, and it informed how Carson, and Rodell, thought about the media response to the book. Although Rodell was the professional in media relations, both had considerable savvy about handling media attention. Meeting the vigorous effort to discredit the book drew heavily on their expertise. In fact, we will see that the very prospect of that effort had much to do with why Carson chose the book form and much to do with how she met the criticisms.

Rachel Carson—Background

Rachel Carson was born in Pennsylvania on May 27, 1907, but her credentials—which were a focal point in the debate about her book—begin with her 1928 graduation with honors in biology from the Pennsylvania College for Women (later Chatham College).[1] She went on to earn a master's degree in zoology from Johns Hopkins in 1931. She taught at both Hopkins and the University of Maryland and spent summers studying at the Woods Hole Marine Biological Laboratory on Cape Cod. She soon took a position with the U.S. Bureau of Fisheries, which later became the Fish and Wildlife Service. Throughout her academic life, her interests had been split between natural science and writing; and as her career in the government agency progressed, she was pleased to find herself combining the two interests, writing articles and short publications for the FWS through the war years. After a brief stint in Chicago in the bureau's war information office, she returned to Washington, where she was made editor in chief of its publications division. She took up residence in suburban Washington, D.C. (Silver Spring, Maryland), where she resided until her death.

During her government work, she was author of record of many FWS publications, but at the same time she pursued her own writing, publishing articles on the ocean, wildlife, and fishing in Baltimore and Richmond newspapers. Her first book, a description of nature in the ocean titled

Rachel Carson, portrait. Yale Collection of American Literature, Beinecke Rare Book and Manuscript Library.

Rachel Carson at work. Yale Collection of American Literature, Beinecke Rare Book and Manuscript Library.

Under the Sea-Wind, was published in 1941 by Simon and Schuster; and she began to have articles published in magazines such as *Nature* and *Collier's* and in Audubon Society publications. With an even fuller, more dramatic discussion of the sea and its life, Carson had her first best-seller in *The Sea Around Us,* published in 1951 by Oxford University Press and excerpted almost simultaneously in the *New Yorker.* Income from *The Sea Around Us* allowed her to resign from Fish and Wildlife to work full-time on her writing, as well as to build a cottage in Maine that became her beloved haven—and in many ways more her home than the Maryland residence.

The Edge of the Sea, published in 1955, brought her to Houghton Mifflin and Editor in Chief Paul Brooks. Her second best-seller, it too was excerpted in the *New Yorker,* this time as the first of the magazine's "Profiles" series not devoted to a human subject. While working on *Silent Spring,* she continued to write articles and book reviews for newspapers and magazines, including *Holiday* and *Woman's Home Companion.* She received many awards for her work even before *Silent Spring,* including the prestigious National Book Award and the Burroughs Medal for distinguished natural history writing, both for *The Sea Around Us.*

Carson had begun to collect material on DDT during the war, and as early as 1955 she had sent memos to Rodell indicating her interest in the effects of pesticides such as dieldrin and parathion.[2] The 1957 lawsuit over aerial spraying on Long Island then drew her intense interest, and her efforts to collect information accelerated. She wrote to Rodell to raise the possibility of writing one or more articles for *Ladies' Home Journal.*[3] At the same time, she wrote to E. B. White, a highly respected author with his own *New Yorker* column, calling his attention to the lawsuit and the attendant issues, and urging him to "take up your own pen against this nonsense—though that is far too mild a word!"[4] White gently declined but did refer *New Yorker* editor William Shawn's attention to the case and to Carson's interest in it.

In concurrent discussions with Shawn and Brooks, plans evolved for a *New Yorker* article and a Houghton Mifflin book for which Carson would act as editor and perhaps contributor of a chapter or two using the same material as that for the *New Yorker.* Shawn's interest in the subject grew and with it, the projected length of the article to a multipart series of twenty thousand to thirty thousand words. Meanwhile, the scope of the topic and the extent of the research facing Carson had made it obvious that help was needed; and Rodell queried several science journalists in search of a collaborator on the book. *Newsweek* science editor Edwin Diamond agreed to work with her on the book for a specified share of the proceeds. However, Carson had no intention of having a collaborator for the *New Yorker* article (nor would Shawn have permitted it), which by that time was so large that it stood in her mind for "the heart of the book."[5] Her vision of the book, therefore, now also called for a single author, who would have to be Carson alone. Diamond parted company with Carson and Houghton Mifflin on somewhat unfriendly terms, which had repercussions in the later public debate.[6] She plunged into the work on her own, thinking it could be ready for publication by summer or fall of 1959.

Unfortunately, the research and writing took more than three years. By itself, the amount of research needed far exceeded her original expectations. But worse, family illnesses and deaths placed heavy burdens on her, and her own health was frequently poor. Delays moved final publication into 1962. The manuscript was completed in January; in June the three *New Yorker* installments were published in the issues for June 16, June 23, and June 30, respectively. Carson, Rodell, Shawn, and Brooks all had hands in editing and revision; and the book was completed by the end of July. Review copies were available the third week in August, and a book reviewers' press conference was held on September 12—the only one she would attend. Official publication date was September 27, 1962. Following publication, Carson was called on to speak to conservationist groups, such as the National Parks Association, and to several influential women's groups, such as the American Association of University Women, the National Council of Women, and the Women's National Press Club. The press sought interviews with her, and she reluctantly permitted *Life* magazine to publish a feature article on her. She was frequently asked to respond publicly to challenges and criticisms of the book. Jay McMullen of *CBS Reports* proposed a report on *Silent Spring*, and Carson agreed, again reluctantly, to participate in a debate for the program, filmed in November 1962 but not aired until April 3, 1963.

Meanwhile, the articles had drawn the attention of President John F. Kennedy, reportedly an avid *New Yorker* reader, who directed his President's Science Advisory Committee to focus inquiry on pesticide abuse. Carson was summoned to speak with its members informally in January of 1963. The PSAC report[7] was issued on May 15, six weeks after the April *CBS Reports* broadcast; and so the first half of the May 15 *CBS Reports* broadcast was devoted to a follow-up, proclaiming that the report had largely vindicated Carson. Shortly thereafter, she was summoned to Congress to testify at Senator Abraham Ribicoff's hearings on environmental hazards and two days later at Senator Warren Magnuson's committee meetings concerning bills on federal spraying regulations. Her public appearances thereafter were fewer and fewer as her health declined. A difficult trip to the West Coast in 1962 had persuaded her to restrict herself to meetings nearer home. Her last major appearance was in early December of 1963 to receive the Audubon Medal from the National Audubon Society. She died on April 14, 1964.

From the beginning of work on *Silent Spring*, Carson had expressed interest in also writing a more positive, encompassing book about balances and nature, introducing the general public to the larger perspective of ecology. She was also working on a book for children that had grown out of a 1956 article for *Woman's Home Companion*.[8] Unfortunately, her health had deteriorated so badly that she was never to see either project come to fruition; but *A Sense of Wonder*, a book combining her text with photographs by Charles Pratt, was published posthumously in 1965. All her books are still in print.

Throughout her life, Carson worked industriously, thoroughly, and doggedly. Her love of nature fueled a tireless love of research, and even though her formal academic training ended with a master's degree—rare enough for a woman of her era—her investigative style was that of a professional scholar and journalist as well. The archives of her papers are dominated by endless boxes of articles and offprints of scientific study reports, correspondence, clippings, and copious notes in numerous notebooks. She was a prodigious reader of books and periodicals, and her notes and letters contain references to most of the major national magazines and newspapers. A mainstay of her research as well as her many friendships was correspondence, and it was common for the letters from friends and acquaintances to include clippings from local papers about issues like conservation, radiation, and the 1959 cranberry scare. Moreover, she never accepted secondhand information but rather pursued a point to its source, communicating directly with a scientist or doctor and obtaining her own copy of the findings. A reference in one article would prompt her to write to another scientist, and a reference in the material she then received would prompt a new round of queries to other sources. As a result, she not only accumulated a wealth of information, but over the course of her career, she also came into contact with a number of experts who became advocates and allies, if not friends as well.

She was, nonetheless, a solitary writer, not a collaborator. She often commented on the loneliness of a writer's work, but the episode with Edwin Diamond illustrated her preferred level of control over her material, both in the collecting and the writing. "Having observed my working habits for a number of years," she wrote Rodell, "you will realize on reflection that 'checking and digging and research' are matters I would never turn over to another person."[9] Where the integrity of her material was concerned, however, she never hesitated to seek the help of colleagues and experts. She sent many of the chapters of *Silent Spring* to colleagues or scientists for verification and comment before submission to Houghton Mifflin, and she welcomed the vetting done by the *New Yorker* and Houghton Mifflin's consultants. Editorial suggestions and revisions from them, from Rodell, and from her publishers were all generally taken with fully professional acceptance.

With the publication of *The Sea Around Us*, she had become something of a literary and naturalist celebrity, drawing attention from those well placed in publishing, the government, and the press. She developed friendships particularly with those concerned about conservation and the preservation of natural resources, including such different people as Interior Secretary Stewart Udall, nature writer Roger Tory Peterson, and Agnes Meyer, wife of Eugene Meyer, owner of the *Washington Post* (Agnes would host a luncheon honoring Carson shortly before the *New Yorker* serialization appeared). A special bond with U.S. Supreme Court Justice William O. Douglas developed after his eloquent dissent from the ruling on the Long Island aerial spraying suit; and he consented to have his endorsement of *Silent Spring*—"the most important chronicle of this century for the human race"—used in Houghton Mifflin promotional material and as cover copy for the Book-of-the-Month Club's booklet announcing *Silent Spring* as its October 1962 selection. Such contacts were to become important supporters for Carson as the public debate mounted.

The success of *The Sea Around Us* had forced Carson to develop public speaking skills, as well as an understanding of the media's handling of celebrity authors. She told a meeting of the American Association of University Women, "I learned the hard way: that people are interested not only in what is between the covers of a book but in the person who put it there. It was naïve of me, I suppose, but when I began to write about the earth and its life, I hadn't realized people would be curious about what sort of woman would write about the sea."[10] Once *Silent Spring* was published, descriptions of her in the press often carried a tone of mild surprise that she was small and "feminine" yet poised and confident, as if journalists expected someone either more imposing or more awkward. As she developed media sophistication, undoubtedly benefiting from Rodell's instincts and experience, she also became increasingly protective of her privacy. Already a reserved person, she was faced with several circumstances that made that privacy particularly important, more so as the *Silent Spring* debate gained heat and she was under mounting pressure to be widely available to the public and the press.

First, Carson's chronically poor health was deteriorating further during preparation of *Silent Spring*. Masses in her breast had been removed in 1946 and 1950; and early spring of 1960, already suffering with an ulcer, she was hospitalized again for removal of new tumors.[11] Initially, she had not been told that malignancy had been found during that surgery; but by the end of the year, with a determined effort to get the truth from a doctor, she learned that she indeed had cancer and that the prognosis was serious enough to warrant radiation treatments. Through the remainder of her life—from 1961 to 1964—she endured several rounds of radiation therapy that made her vulnerable to ancillary ailments and opportunistic viruses. Added to the nausea and extreme weakness from the radiation were other stomach problems, bronchitis, an eye inflammation that rendered her temporarily blind during a critical phase of the book, and knee and heart problems that hampered her productivity and eventually curtailed her public appearances.

To say that it was a wonder that she was able to complete her work at all is an understatement, yet many have reported that she acquired a sharper sense of urgency and purpose as a result of her physical afflictions. Wary of interpretations attributing a bitter or vindictive tinge to her reasons for writing about possible carcinogens in chemical pesticides, Brooks made a point of stressing that she learned of her cancer long after she was well into her work on the book.[12] Her correspondence does include requests for information on chemical links to carcinogenesis dated well before she got her grim prognosis in late 1960; the specter of cancer had likely hovered from the earliest surgeries. Nonetheless, she was well aware that her critics, had they known of her cancer, could easily have claimed that the diagnosis had prompted her campaign and that her work was an irrational attempt at vengeance against imaginary culprits.[13] For that reason alone, it is easy to understand her desire for strict secrecy about her condition right up to her death—only Rodell, Brooks, and at the last a very few intimate friends knew.

But she had other reasons for maintaining extraordinary privacy. Her family situation was complicated and potentially embarrassing. Her mother, a niece, and the niece's out-of-wedlock child were living with and dependent on Carson. The niece died suddenly in early 1957; and as work on *Silent Spring* began, Carson found herself caretaker of both an aging mother and very young grandnephew. Public knowledge of the child's background in the 1950s would have been shameful for a family to a degree unthinkable in today's culture; and although Carson had satisfactorily explained the absence of a father to most,[14] she could never be sure the strong light of celebrity might not expose and somehow exploit the situation. Possibly more compelling, Carson herself had close and often intense relationships with her friends; and much emotional intimacy, particularly with her closest longtime friend, Dorothy Freeman, is evident in their correspondence (not published until 1994).[15] Again, insinuations about the nature and quality of such close relationships between women would have been—in the atmosphere of the era—so taboo that even hinting about it could have meant wholesale condemnation, although a couple of her detractors seemed tempted to do just that.[16] As it was, the fact of her gender was so much an undercurrent in the *Silent Spring* debate that Carson was undoubtedly on the alert for any spin that issue might have taken.[17] Finally, her celebrity with the success of *The Sea Around Us* had given her ample experience of the trials and pitfalls of even limited notoriety, which she found profoundly distasteful. In fact, she welcomed contractual limitations on her public and media appearances in 1962 and 1963. At the same time, her unavailability may actually have enhanced her status in the public eye.

Rodell had expertise in public relations that allowed her to protect her friend even while promoting her client, although the word "client" was probably never used between Rodell and Carson. "Five years younger than Carson," as Carson's biographer, Linda Lear, described her, "Marie Rodell was, by 1948 [when they met], a world-wise, well-traveled, sophisticated New Yorker fluent in four languages who moved comfortably in many of New York's most elite literary and publishing circles. . . . A member of MENSA, the society for people with extremely high IQs, Rodell spent her spare time writing mystery fiction and published three novels."[18] Educated at Vassar, Rodell began her career as assistant editor in a book publishing house, moving on to others and doing some writing of her own, too. In 1948 she undertook to become an independent literary agent, and through a mutual publicist friend she met Carson, then in the process of changing publishers. Thereafter her professional life was increasingly dominated by her work with and for Carson.

By midcentury, literary agents had come to be far more than soliciting intermediaries between author and publisher. Increasingly, they were involved throughout the process of writing and editing, publication planning, and ultimately promotion. Such agents were therefore more than highly literate advisers on publishing matters. They had to be well-networked manipulators of opportunity and media, who knew the critical people to get notice for a book—not only editors, publishers, and literati but book reviewers, journalists, and celebrities as well—with a precise sense of timing. Above all in the media-dominated twentieth century, they had to be masters of the art and science of publicity, looking out for the best interests of book and author, and with or without a good working relationship with the publisher's own publicity and marketing departments. Rodell's skill in negotiation was more than matched by her sensibilities regarding the placement and timing of promotional efforts. In addition, her New York connections and experience likely gave her an advantage over those based in Houghton Mifflin's Boston headquarters. Despite their enthusiasm for the book, Rodell sometimes felt that they seriously underestimated its potential impact;[19] and her relationship with Houghton Mifflin (particularly Executive Vice President Lovell Thompson) was not always smooth. She frequently stepped in when she felt their efforts were falling short, and occasionally some of her impatience with the Houghton Mifflin publicity department surfaced despite an earnest effort to maintain good relations on the part of both Rodell and Anne Ford, head of Houghton Mifflin publicity.

At the same time, Rodell became increasingly protective of Carson as her health failed, fielding attempts to draw Carson back into the spotlight without giving away the seriousness of her condition. Before Carson's death, the two began to go through Carson's papers in preparation for donating them to Yale, and Rodell later devoted two years to completing that task. Moreover, Rodell pursued various means for continuation of Carson's work after her death, including the posthumous publication of *A Sense of Wonder*, because she had come to share Carson's mission.

The Mission

"What has already silenced the voices of spring in countless towns in America? This book is an attempt to explain."[20] Rachel Carson thus told the readers of *Silent Spring* that, first and foremost, her intent was to *inform* them about something: the damage wrought by pesticides applied without thought to their hazards. That purpose had been the core of her work from the beginning. When she wrote E. B. White early in 1958 to suggest he cover the Long Island aerial spraying trial, she began, "In recent weeks I have been reminded of my own former conviction that the mass spraying of DDT and other even more dangerous insecticides is a threat to the entire balance of nature and even more immediately to the welfare of the human population." She mentioned several reports from credible sources and noted plans for future spraying, saying, "There is an enormous body of fact waiting to support anyone who will speak out to the public."[21]

Having decided to take on the topic completely herself, by June of 1958 she was feeling an increasing sense of urgency, writing to Dorothy Freeman, who already feared the probable controversy: "You do not know, I think, how deeply I believe in the importance of what I am doing. Knowing what I do, there would be no future peace for me if I kept silent. . . . [I]t is, in the deepest sense, a privilege as well as a duty to have the opportunity to speak out—to many thousands of people—on something so important."[22] As time went on, she noted that while she was not the first to write about the problem, the general public was still ignorant: "The book is desperately needed. Unquestionably, what it has to say will come as news to 99 out of 100 people."[23] Later she expressed her belief that the importance of the topic itself outweighed her role in illuminating it: "If I had not written the book I am sure these ideas would have found another outlet. But knowing the facts as I did, I could not rest until I had brought them to public attention."[24]

Carson, second, sought to expand awareness to the broader question of overall human behavior within the natural system. She explicitly subscribed to Albert Schweitzer's "reverence for life," and her first three books were literary exemplars of the philosophy. She sought to infect her readers with a love and understanding of nature, which Carson conceived of as an intradependent system. But for Carson, the complement to Schweitzer's reverence was Rostand's right of the people to know. With other naturalists, she could use the phrase "balance of nature" expecting to be well understood, but its use in mainstream discussion at the time almost guaranteed that the user would be considered a fringe-group mystic. Although her earlier books had carried no admonitory tone, her 1961 reissue of *The Sea Around Us* included a new introduction warning that disposal of atomic wastes into the ocean could have catastrophic consequences over time, as containers failed and released the wastes: "The sea, though changed in a sinister way, will continue to exist; the threat is rather to life itself."[25]

Silent Spring was thus intended to do far more than provide understanding and inspire reverence. In fact, the original working title for *Silent Spring* was "Man Against Nature," and its dedication to Albert Schweitzer is accompanied by his ominous words: "Man has lost the capacity to foresee and to forestall. He will end by destroying the earth."[26] Not long after the *Silent Spring* controversy first reached national view, Carson began to make a point of saying publicly that pesticide misuse was only part of a larger problem with human stewardship in the natural world. She told the *New York Times Book Review*, "The problem I dealt with in *Silent Spring* is not an isolated one. The excessive and ill-advised use of chemical pesticides is merely one part of a sorry whole—the reckless pollution of our living world with harmful and dangerous substances."[27] On Capitol Hill in 1963, she told the Ribicoff committee, "The contamination of the environment with harmful substances is one of the major problems of modern life. . . . The problem of pesticides can be properly understood only in context, as part of the general introduction of harmful substances into the environment."[28]

The more urgent she felt about informing the public, the more a third aspect of her mission came into play: not only did the public have a need and a right to know, but public attitudes and actions also needed to be mobilized. The right of the public to know was predicated on the idea that self-governing action followed from full information. Shortly after the book was published, she was to note that hitherto, the public had seemed "serenely unaware and unconcerned";[29] but the reason for their ignorance was not apathy but rather having been deliberately kept in the dark by a "vacuum of secrecy."[30] Since part of her mission was breaking that vacuum, she would supply readers of *Silent Spring* with the facts pertinent to needed action. She asked them, "When will the public become sufficiently aware of the facts and demand such action?"[31] During the *CBS Reports* follow-up program, Carson said, "As [the presidential] panel makes clear, there are decisions to be made. The public can make them only when in full possession of the facts; and I am particularly pleased by the reiteration of the fact that the public is entitled to the facts, which after all, was my reason for writing *Silent Spring*."[32]

For her part, Rodell's enthusiastic plan for early promotion included distribution of proof copies, as soon as available, to major policymakers and national political leaders, not only to solicit "publishable statements" of endorsement but also to begin the public discussion among those in a position to affect change.[33] For Carson, however, even though she had worked in the Kennedy campaign in 1960, direct and effective interaction with government depended on having a popular mandate to do so.[34] What she wanted of the public involved both supervision and legislation, and four months after publication, she observed, "People are beginning to ask questions and to insist upon proper answers instead of meekly acquiescing in whatever spraying programs are proposed. . . . There is an increasing demand for better legislative control of pesticides."[35] Carson thus framed her role in the debate as mobilizing messenger, carrying the alarm to the people, who would themselves then demand greater responsibility, more information, new legislation, enforcement, or other voluntary controls, from those in government and industry—not to mention modifying their own behavior as pesticide users. In these respects, Carson's mission was very much that of a journalist's, even if her advocacy was conveyed in book form rather than in the traditional form of a newspaper editorial.

Eventually, a hopeful theme arose in her accounts of public reaction to *Silent Spring*. Her interpretation of her audience's urge to act was that it represented "the reappearance of a sense of personal responsibility."[36] That "personal responsibility" was not just a matter of monitoring one's own decisions but of taking responsibility for community decisions rather than waiting for the government to provide information and to take action on behalf of the populace. "Trusting so-called authority is not enough. A sense of personal responsibility is what we desperately need," she told *Life*.[37] For *New York Times Book Review* readers she wrote, "Until very recently, the average citizen assumed that 'Someone' was looking after these matters and that some little-understood but confidently relied-upon safeguards stood like shields between his person and any harm. . . . [T]he public [now] understands that these problems do not correct themselves."[38]

Carson's understanding of the reasons for such governmental irresponsibility underlay a fourth, less explicit but nonetheless compelling, aspect of her mission in writing *Silent Spring*, that of, in effect, blowing the whistle on the political and economic processes that had either created the problems or permitted them to continue and even worsen. While she was a federal employee within Fish and Wildlife, dissent or criticism of federal programs would have been, to say the least, discouraged. Early on in her work, there were indications of her dim view of her former employer: "Someday I wish I could find time to turn my pen against the Fish and Wildlife Service's despicable poisoning activities! I do think I can work in references to this sort of thing in the present book—it is all part of the same black picture."[39] In fact, a passage in her manuscript for *Silent Spring* described retaliatory pressure against whistle-blowers extending beyond Washington, to which she called Brooks's attention: "A statement at the bottom of p. 219 to the effect that the Department [of Agriculture] tried to have various state and federal biologists fired may have to come out. I know it is true and I am now trying to get copies of letters that prove it. Unless I can get them, however, I will drop it."[40] The passage was dropped. But as a private, self-employed citizen, Carson the author remained free to use media at her disposal to criticize governmental programs as she might choose.

Incensed particularly by the fire ant and gypsy moth programs, she specifically targeted the Department of Agriculture (USDA), although other departments within Interior, including her former division of Fish and Wildlife, drew fire as well. She explicitly took on the USDA's aerial spraying programs in chapter 10 of *Silent Spring*, "Indiscriminately from the Skies," a passionate but well-documented criticism of the "eradication" programs. Those programs were being conducted by the USDA despite compelling evidence that the programs—aimed at complete elimination of a given type of insect or pest—would not and could not work; that the intended targets posed comparatively little threat to begin with; and that the collateral damage of sprayed chemicals to humans, pets, livestock, and natural resources was real and serious.

The reason for the USDA's insistence on such programs and such flagrant dismissal of contraindications was, as Carson understood it, the close relationship between chemical companies and government agencies. Because the direction of Agriculture, in particular, came largely from agribusiness interests, its sympathy with producers of agricultural chemicals meant that it relied on scientific studies of pesticides funded by the same companies. And funding for academic studies of pesticides was also coming heavily from chemical companies interested in protecting their products; thus academic researchers concentrated on efficacy rather than on hazards and side effects. Researchers knew the chemicals killed pests; they seemed uninterested in what else the chemicals did. Inevitably, findings favorable to greater sales and use of chemicals would color inquiry.

For Carson, it was an unholy, and not entirely voluntary, three-way alliance among the chemical industry, government, and academic science. As early as 1953 in a letter to the *Washington Post* deploring the dismissal of Alfred M. Day as director of Fish and Wildlife by an Eisenhower appointee known for his contempt of conservationists (he was said to have characterized them as "punks"[41]), her concern for the government's environmental policies was evident. The administration of the nation's natural resources "is not properly, and cannot be, a matter of politics," she wrote, concluding that the "hard-won progress" of public-spirited workers to preserve natural resources was "to be wiped out, as a politically minded Administration returns us to the dark ages of unrestrained exploitation and destruction."[42]

In addition to an awakened, properly skeptical public, another remedy for the imbalance in pesticide policy and administration would therefore be to restore supervisory authority to those without vested interests. Her appearance before the Ribicoff committee on government reorganization focused particularly on the need to balance the power among government agencies concerning pesticide policy. She testified, "There may easily be serious conflicts of interests between such varied segments of our economy as agriculture and the commercial fisheries. It seems to me that there should be no automatic assumption that the agricultural needs should be served without regard to damage to fisheries; or, in other situations, to wildlife. . . . This matter of conflicting interests, and of conflicting governmental mandates, lies at the heart of the problem this legislation is designed to solve."[43] A shoring up of the power of the Food and Drug Administration (FDA)—theoretically less vulnerable to pressure from the agricultural chemical industry—to police contamination and to enforce limits on it would also have shifted the balance back toward less vested interests. In a somewhat complicated discussion of FDA "tolerances"— permissible limits of contamination—Carson pointed out that "to establish tolerances is to authorize contamination of public food . . . and then penalize the consumer by taxing him to maintain a policing agency. . . . So in the end the luckless consumer pays his taxes but gets his poisons nonetheless."[44]

The PSAC report addressed this issue directly; and in her published response to it, Carson pointedly mentioned her happiness that the panel had called for a review of tolerance policy and policing. Overall, after an era in which "the insect control agencies seemed to be in the saddle," Carson found hope in the PSAC report that "control programs will be put into perspective against other interests involved." Specifically, supervision of the chicken coop needed to be wrested from the fox: "The Federal pest control review board is made up of representatives of the very agencies in pest control, and so, in effect, has been asked to pass judgment in its own actions. While it has occasionally modified a program, it has never recommended that one be discontinued."[45]

As much as the system of interlocking vested interests appalled her, it was the effort to lull the public into unquestioning acceptance of pesticide use, including aerial spraying programs, that provoked her greatest ire. Sophisticated enough to know a public-relations campaign when she saw one, she was much annoyed by the USDA's campaign to reverse negative attitudes toward the fire-ant spraying program, as typified by a widely circulated public service film, *The Fire Ant on Trial,* that particularly incensed Carson.[46] In chapter 10 she wrote, "In 1957 the United States Department of Agriculture launched one of the most remarkable publicity campaigns in its history. The fire ant suddenly became the target of a barrage of government releases, motion pictures, and government-inspired stories portraying it as a despoiler of southern agriculture and a killer of birds, livestock, and man."[47] A few paragraphs later, the word "propaganda" is used to refer to material used in this public-relations effort, and the term appears frequently thereafter in the book. Similarly, she deplored the failure of government and chemical companies to educate consumers about their own use of pesticides through cautions in packaging and advertising. "Lulled by the soft sell and the hidden persuader, the average citizen is seldom aware of the deadly materials with which he is surrounding himself," she wrote.[48] Thus, part of the book's role as tocsin was not only to awaken the public to the dangers of pesticides but also to the dangers of passivity in the face of government and chemical company "propaganda."

Overall, Carson's mission in writing *Silent Spring* had four elements: first, to awaken and to inform the public about the hazards of chemical pesticide use; second, to increase awareness of the larger ecological problems facing the earth; third, to mobilize the public to effect changes in policy and practice; and fourth, to challenge the political and economic relationship between industry, science, and government—particularly the USDA. Oddly, Carson was often at pains to deny that she was a crusader, yet the sum total of her objectives in writing *Silent Spring* certainly had the effect of a crusade. It never occurred to her, for example, to include any discussion of the historical benefits of chemical pesticides—something her opponents thought she owed them—just to create a balanced presentation. Though she fully expected a negative response, the evidence suggests that the scale of the public relations war surprised her a bit. Her surprise is partly explained by the probability that she did not see her work as a unilateral, gratuitous attack on those she criticized but rather as a correction of an egregious imbalance in information, the distorted result of an ongoing public-relations effort to lull the public into uncritical acceptance. Perhaps for that reason she did not embrace the image of crusader—someone initiating civil assault for the purpose of effecting wholesale change—so much as that of monitor and advocate, even journalist: someone who perceived that authorities had strayed from acting in the public interest and who advocated restoration of proper regard for the public in policy and behavior.

Carson and Her Audience

In the shift from nature writer to advocate, Carson's relationship with her readership necessarily changed. As an author, she had always, of course, assumed the existence of a reader with whom she shared the traditional, dyadic author-reader relationship and whose understandings and needs she was obliged to take into account. The response of that reader might be entirely interior or, at most, shared personally with the author or another single reader through letters or conversation. Her fourfold mission for *Silent Spring*, however, called for a more collective idea of readership: an audience consisting not only of an expanded number of people but also of a community no longer limited to nature lovers and armchair biologists. As all authors do, she communicated to form a particular kind of connection with those who read her book or the excerpted form; but part of her strategy in pursuing her mission was to use that positive, individual connection itself to command a stronger relationship with the media at large, thereby enabling her to communicate with a still greater audience. Her challenge was not only to explicate difficult material but to do it so well that, on the basis of what she wrote, members of her audience could be moved to take committed civic action, even if there were public criticism of her message. Considerable study already exists concerning the specifically rhetorical aspects of *Silent Spring* and the debate around it,[49] but here it is important to note how awareness of the audience conditioned presentation of Carson's message, particularly with respect to media participation in the debate.

To begin with, *Silent Spring*'s readers would be required to follow the physical, chemical, and biological intricacies, as well as the chains of causality and consequence, in order to understand, for example, why water pollution could never be confined to one geographical location. "How to reveal enough to give understanding of the most serious effects of the chemicals without being technical, how to simplify without error—these have been problems of rather monumental proportions."[50] The size of the task indicated using the book form, yet by its very dimensions such a technical volume could be intimidating. Carson's use of the fable format to ease her readers into the book was an inspired way of establishing why all the science mattered to everyone. Thereafter, the challenge was to keep the reader engaged in the technicalities. "Since this really basic process [energy-production through phosphorylation] is disrupted by some of these chemicals, I have to bring to life for my readers what goes on inside the cell and even inside the mitochondria."[51]

Part of Carson's strategy also relied on awareness of what the general public already knew about environmental dangers. The comparison of chemical toxicity to radiation poisoning was an intentionally recurrent theme in *Silent Spring*, not only providing a useful parallel to facilitate comprehension but also tapping into public anxieties about nuclear fallout. "We are rightly appalled by the genetic effects of radiation; how then, can we be indifferent to the same effect in chemicals that we disseminate widely in our environment?"[52] The 1959 cranberry scare had also caught Carson's rapt attention (she attended a congressional hearing on the FDA ban of possibly contaminated cranberries[53]). She often suggested that popular acceptance of *Silent Spring* was facilitated by the time and general awareness of that scare together with others involving milk contaminated with strontium-90 and birth defects caused by the drug thalidomide.

Inevitably, her audience would include those undaunted by the technicality but hostile to the book's message itself, among them the very scientists and chemical company officials she faulted for the situation. Thus, her words had to be carefully chosen. Marie Rodell, along with the *New Yorker*'s William Shawn and Paul Brooks, worked with her to modulate the levels of technical language and potentially inflammatory language. At one point Rodell wrote, "In one place you refer to the 'enormously profitable' chemical industry and its reluctance to give up its profits. If this is instead phrased in their own language—in what would be part of their retort—it takes the wind out of their sails—and also spreads the blame to all who share it."[54] Consciousness of an unfriendly, oppositional audience was essential to the process of writing and editing throughout, even if the size of a potentially receptive audience would be far greater; her opposition would have full access to the media, Carson knew.

Despite Carson's broad view of her audience, in some important ways her image of her readership remained not in the collective but rather quite personal. She considered each of her many professional and personal correspondents her reader, as they almost always were; and over her career she received many letters from individual readers among the general public—

some of whom joined her list of regular correspondents. Until the volume of letters responding to *Silent Spring* became overwhelming and her energy began to wane, she made an effort to respond personally to the letter writers. Indeed, the legendary stimulus for writing *Silent Spring* was a 1958 letter describing some songbirds' death following a spraying of DDT. Carson pointedly cited Olga Huckins's letter publicly and privately as the impetus for the *Silent Spring* project,[55] even though the two were already acquainted and Carson's concern about pesticides, DDT particularly, long predated the Huckins letter.

For Carson, there was considerable crossover and interrelation among letters to her, letters to editors, and her public appearances. She quoted letters to audiences at speeches, she referred to letters in interviews, and she referred to press coverage in letters. She drew considerable personal strength, encouragement, and gratification from the personal contact—especially when the letters reported actions that proved her mobilizing mission had been accomplished. Perhaps more significantly, reporting that kind of evidence of her success and public support was an important part of her post-publication strategy for defending her message.

Despite her particular antagonism toward spraying programs coordinated at the federal level, her concept of appropriate civic action began at the local level; and many of her speeches cited letters about local efforts by individuals and groups in response to *Silent Spring*. One Connecticut science professor wrote her that he had helped set up an impartial committee to review state practices, which struck her as the perfect example of the kind of action she had in mind: "I am, in fact, calling attention to your group and its activities as a sort of model when people write me . . . about the possibility of forming a national organization. I am convinced that state, or even smaller regional organizations are much more effective and can retain better control of what the group stands for."[56] Nonetheless, the overall scope of her mission was national. In late spring of 1963, the near simultaneity of the release of the PSAC report and her appearance at congressional pesticide hearings indicated how successfully she had gotten the attention of both executive and legislative branches of government, such that both branches were compelled to take some public sort of action. Insofar as DDT would eventually be banned and more scrutiny would fall upon practices involving other pesticides within a few years after her death, her primary mission in some measure was accomplished.

Yet notably more gratifying was the fact that it was outcry from individual members of the public, not her personal appeal to government, that prompted the governmental response. "Most heartening of all are the number of people who want to *do* something. And this is a personal, an active thing. . . . They ask, 'what can *I* do to help?' "[57] She saw herself as facilitator of the process, not its architect: "I came to realize that scattered throughout the country were thousands of people who were concerned—who were trying, as individuals or as small groups, to do what they could, in the face of great odds. Now, simply because I happen to have brought together the basic facts—because I have written a book that seems to be serving as a rallying point for an awakened public—both the strength and the needs of these people are flowing to me in a vast and wonderful way."[58]

Handling the Media Storm

The single act of writing a book by no means guaranteed that it would become "a rallying point for an awakened public." Once published, the book still had to come to the attention of the public before the public could be awakened by its content. Beyond that, the rallying process would entail knowledge and skill in dealing with public opinion processes, those occurring within the media above all. It behooved Carson and Rodell to play their part not only in promotion of the book but also in its defense in the public arena once the controversy gathered steam.

Undoubtedly, they were already prepared to do for *Silent Spring* what would have been asked of any author and agent to publicize any book: get word to book reviewers, make sure of the publisher's advertising to booksellers, capitalize on connections in the media, and be prepared to participate in book parties or signings in key urban bookstores. As far as possible under the circumstances of Carson's failing health, they were indeed concerned with the "usual" business of selling a book. But their commitment to their mission concerning pesticide hazards had intensified the desire for the broadest possible exposure, even before Carson was called on to answer her critics publicly. Delighted to hear that the book had been made a Book-of-the-Month Club selection, Carson said, "No one could say whether total sales and income will be greater this way but what gives me deep satisfaction is the feeling that this . . . will give it an irresistible initial momentum. And the BOM will carry it to farms and hamlets all over the country that don't know what a bookstore looks like—much less the *New Yorker*."[59] Her enthusiasm suggests strongly that Carson was not necessarily intent on getting her book (or the *New Yorker* series) purchased and read but rather on achieving "momentum" for the message it carried.

When *Reader's Digest* disappointingly dropped plans for a condensation of the book, Carson wrote, "With that 'mass medium' eliminated, everyone felt I should take advantage of *Life*'s pages. Well I hope I won't regret it."[60] Carson had not wanted to have anything to do with *Life* after unpleasant dealings in 1956 over an article on the jet stream that she had ultimately declined to do. Moreover, she disliked the personal tone of the magazine's articles on celebrities; but eventually she did agree to the article, especially after Rodell pointed out "that if [*Life* publisher] Henry Luce opened up his pages to the chemical industry, it would certainly seize the opportunity."[61]

Despite intermittent illness and mounting fatigue, Carson consented to a number of speeches at garden, nature, and women's clubs. Although this was a speaking circuit that, at first, very likely paralleled those she had traveled with *The Edge of the Sea* and *The Sea Around Us*, these events afforded opportunities for press coverage, and in a more controlled situation than the press conference format, which she avoided on all but one occasion. In lieu of subjecting herself to interviews, these addresses had to be a primary part of her relationship with the media, with the crucial advantage that they demonstrated immediate public interest in what she had to say. When the Garden Club of America declined to notify the press of Carson's speech there, Rodell wrote a note of exasperation with the club to Houghton Mifflin's publicist.[62] Ever vigilant, Rodell monitored the output of clipping services and kept copious records of the number, sources, and types of mentions of Carson and the book—carefully listing articles, columns, reviews, and even letters to the editor, until the volume became overwhelming.

The prospect of a public challenge to *Silent Spring*, however, demanded alterations in the usual promotional strategies. The risk at hand was that *Silent Spring*—both its message and Carson herself—might be so discredited that their message would not be received—well or at all—by the general public. Although they expected negative response from those Carson criticized, the manner and degree of that response in the public forum could not be predicted. A strategy of risk management, so to speak, was inherent in how Carson and Rodell approached the media, at a time when the term "risk" was scarcely part of an author's or agent's standard professional vocabulary. They would need the media to disseminate, promote, and defend Carson's message; but Carson had some experience in the pitfalls of media exposure, and she and Rodell knew that the most serious threats to the viability of *Silent Spring*'s message would also come through the media.

One way of dealing with media-borne attacks on book and author was to control exposure. Two circumstances imposed welcome limitations on the timing and degree of public exposure: Carson's deteriorating health and the terms of their agreement with CBS for the *CBS Reports* program. Already exhausted at the end of a summer of lively controversy over the *New Yorker* articles and prepublication activity at Houghton Mifflin, Carson was secretly undergoing radiation therapy during the fall after publication. Her public appearances were sparse. In consultation with Rodell, she agreed to a book reviewers' press conference and a cocktail party whose guest list was confined to book-page journalists and scientists, instead of a much more elaborate Houghton Mifflin book party.[63] She attended a reception and book signing in Cleveland, where she said she had close professional ties but where in fact her preferred doctor had his practice. But other than her speeches to the National Audubon Society, a few garden clubs, and the National Women's Press Club, she was rarely available to the media.

Radio and television broadcasters quickly sought her out, but when she agreed to do the *CBS Reports* program with Eric Sevareid, she had signed a contract stipulating that she do no other programs until three months after its broadcast. Although taped in November, the program was not aired until April, which meant that for five months Rodell could turn down other interviews with no further explanation other than reference to the CBS contract, which suited Carson perfectly.[64] As it was, she was apprehensive about the program ("I'm pretty sure I won't like it") and about the roster of other interviewees, expressing to her friend Dorothy Freeman that "it seems to me the show is weighted against me, and I'm rather annoyed."[65] When the broadcast date was announced five months later, she fretted about how she would come across, especially with her extreme exhaustion and husky voice from the medical treatments at the time; "I just hope I don't look and sound like an utter idiot."[66] Viewed decades later with an informed eye, her appearance does gives slight hints of her condition (her hair could well have been a wig), but her composure in contrast to the heat of her primary challenger, Dr. Robert White-Stevens, worked much to her advantage; and most contemporary newspaper accounts of the program emphasized her poise and calm.

Beyond Carson's reluctance to breach her own privacy and subject herself to fatiguing public interviews and debate, Carson and Rodell saw other strategic reasons to limit her exposure. Early in her work on the book, it was clear that the chemical companies had got wind of her project, and she and Rodell quickly saw that foreknowledge might give her challengers

Carson with CBS's Eric Sevareid in her Silver Spring home. CBS Photo Archive.

too much advantage. Unavoidably, her own queries to government offices and laboratories had given some indication of the direction of her work, and occasionally she received responses questioning her authorization or purpose. One reason for asking Edwin Diamond to work with her had been that he might have better access to certain such sources. The large number of her correspondents also made her vulnerable to leaks about her project. As early as November 1958, she wrote to her friend and former FWS colleague Dr. Clarence Cottam, "The news seems to be out on the grapevine."[67] A few months later she wrote him again: "I feel that premature disclosure of even part of this material would do more harm than good to our cause. . . . The whole thing is so explosive, and the pressures on the other side so powerful and enormous, that I feel it far wiser to keep my own council [*sic*] insofar as I can until I am ready to launch my attack as a whole."[68]

Three preliminary public tip-offs about her work had demonstrated the risks of premature notice. In April 1959, still early in her work, she wrote a letter in response to a *Washington Post* editorial concerning the death of migratory birds after a harsh winter, in which she described the "sudden silencing of the song of birds" by chemical pesticides.[69] She was flooded with supportive responses, and commentary appeared in several newspapers. Next, Houghton Mifflin itself unwittingly prompted notice by prematurely listing the book in its summer 1959 catalogue (having assumed Carson would complete the work within a few months). When Carson's mail brought a "sudden rush of clippings" about the new book, she quickly had Rodell tell Paul Brooks that all future copy about the book would have to be submitted to Carson for review before release, since "the intent of the book can be distorted innocently but badly with only slight differences in wording."[70] Finally, in an unusual underestimation of the interconnectedness of the media system, Carson wrote a letter to the *Boothbay (Maine) Register* in July of 1961 commenting on an article on tree spraying. It was picked up by wire services and shortly appeared around the country. She wrote an Audubon Society administrator, who had wired her asking permission to reprint the letter, "Perhaps I should never have written the letter to the *Boothbay Register* in the first place because I am really not ready to have widespread publicity given to any of my ammunition. To write as I did for a small and local paper seemed deceptively harmless and I should have known better."[71]

Her worries that her message might be distorted before it even appeared in her own printed words joined a larger concern about credibility—the book's and her own as its author. At the heart of every decision regarding the media and the book's critics was Carson's and Rodell's determination to protect credibility above all else. They knew from the outset that the opposition would devote considerable energy to undermining the public's estimation of the author, her credentials, and her evidence. Both Carson and Rodell were faced with anticipating all forms of personal attack in addition to attacks on the book's content, and they needed to devise effective ways of deflecting or neutralizing the attacks they knew would come from many quarters, even if, in their own minds, the financial interest of the chemical industry was behind it all.

First, challenges to Carson's credibility as author would involve her credentials as a scientist, her motivation in writing the book, and her rationality with respect to her subject. Rodell made sure that Houghton Mifflin noted Carson's degrees and professional scientific work on the book's jacket copy and that publicity materials from Houghton Mifflin included similar references. Rodell took severe note of anyone understating or denigrating the depth of Carson's knowledge, writing any offending editors to demand correction. When in late 1962 Houghton Mifflin decided to issue a promotional brochure to respond to the various attacks and criticisms, Carson insisted on considerable involvement in the writing of it; and much of her attention was directed to what she felt was a crucial section titled "Is the Author Qualified?"[72] Presentation of her credentials was a significant part of each public appearance, not the least before the June 1963 Senate hearings, where she also reminded those in government that, although a private citizen, she had considerable firsthand knowledge of government processes: "I speak not as an outsider but as one who has had some 16 years' experience as a Government biologist. I, therefore, am well aware of the problems, the frustrations, the inevitable conflicts that arise when two or more agencies attempt to carry out their sometimes conflicting mandates."[73]

Dealing with subtler efforts to undermine her character was more difficult than defending her background. Aware that even Carson's illness offered potential vulnerability if critics could imply that she was motivated by bitterness because of her cancer (a disease linked to toxic chemicals in her book), Rodell was careful to monitor the tone of what Carson wrote, at one point writing Carson, "The references to and comments about the chemical companies need to be phrased very carefully not to set up a counter-reaction in the readers. We must at all costs avoid giving anyone the opportunity to yell 'crank.' "[74]

Her gender was another point of potential vulnerability,[75] at least in the minds of the opposition and some of the media, who occasionally seemed to put excessive emphasis on the honorific "Miss" in her name (as opposed to "Dr." or "Mrs."). Former Agriculture Secretary Ezra Taft Benson was said to have asked why "a spinster with no children is worried about genetics,"[76] although his choice of the word "spinster" courts speculation about a nastier subtext than plain misogyny, given the times. That Carson might have been considered "only" a woman's author, however, is perhaps less surprising in light of the fact that, except for the *New Yorker*, she had been writing articles most often for women's magazines and spoke most frequently to women's clubs or to clubs with predominantly female membership. By the time she was called on to speak about *Silent Spring*, she began to make a point of noting how much of her audience and how much of her popular support had come from men: "57 percent of those writing were *men*, which pretty well disposes of the legend that it is only 'the ladies' who are concerned about pesticides."[77]

A few even used the term "Communist" against her (Benson speculated that Communist leanings might somehow explain her interest in genetics), but those who did so employed conflicting logic. Some faulted her for unpatriotic criticism of the government, others faulted her for seeking excessive federal involvement. Nonetheless, wary of giving any opening to her opposition, she and Rodell agreed that there should be no sales of foreign literary rights to Iron Curtain countries "because the book is too easily twisted to anti-U.S. propaganda."[78] In an era still mired in the Cold War, the word "propaganda" itself, which Carson and her critics used against each other, carried heavy connotations of enemy subversion.

Both Carson and Rodell worked assiduously to dissociate Carson from any fringe groups, particularly those implying that she endorsed a particular product or practice. Both felt they needed to avoid any taint of irrationality or "hysteria"—a word used constantly against *Silent Spring*. But more important, if there were any suggestion that Carson's first priority was to make money, or worse, that she was funded by anyone other than herself, her own criticism of the financial dynamics of the pesticide problem would have been rendered hypocritically hollow. She and Rodell were so chary of the appearance of endorsement or association with interest groups that they insisted that all printings after the first include a disclaimer paragraph at the beginning of the book, prohibiting the use of Carson's name or the book's title to promote any product.[79]

Her greatest strength and protection, however, was that the book rested not on her word but on the authority of all those sources on which she had drawn. In derisory response to a chemical journal reviewer who called her reference citations little more than "name-dropping," she told the journalists at the Women's National Press Club, "Well, times have certainly changed since I received my training in the scientific method at Johns Hopkins! . . . Now I would like to say that in *Silent Spring* I have never asked the reader to take my word. I have given him a very clear indication of my sources. I make it possible for him—indeed I invite him—to go beyond what I report and get the full picture. This is the reason for the fifty-five pages of references."[80]

Those fifty-five pages were the foundation on which Carson's position rested, and Carson and her supporters could point to them repeatedly as evidence of the solid science behind her report. Her confidence in her sources came, moreover, not only from the quality of their work but also from the fact that everything she wrote had been vetted along the way by still other scientific experts. For added protection, she and Rodell had made certain that what she wrote was vetted by Houghton Mifflin lawyers for libel, taking great care to note which chemical names were generic and which proprietary.[81] Any actionable bit of carelessness would give chemical companies not only leverage to impede publication but additional grounds for impugning her methods.

To the documentary support of her argument Carson and Rodell could add the professional and personal support of many public figures, as well as the approval of the public at large. Rodell's prepublication efforts to place proof copies in the hands of policymakers and opinion leaders was part of a conscious effort to have heavy-hitter support already in place when the book was published. She wrote Anne Ford, "Since Rachel is undoubtedly going to be attacked from some quarters as a crackpot and subversive, a back-log of highly respectable people who have read the book and discussed it with her will be an enormous help."[82] Rodell and Carson worked with Houghton Mifflin to select supporting statements from significant scientists as well as from respected figures in government (particularly Justice Douglas) for publicity and advertising copy. In her speeches, Carson typically made note not only of expert support and corroboration but also of popular support for her cause, listing the number and tone of letters and reading from some of them to give her audience the flavor of the public's endorsement. By the time of the release of the PSAC report, she was concentrating on that popular support as, by implication, the most valuable of all.

Once the full force of attacks on Carson and *Silent Spring* began to be felt, Carson moved not only to answer the criticism but to go on the counterattack as well. The opposition tended to overstate her case—saying that she favored wholesale discontinuation of all pesticides, leaving people, agriculture, and countryside to the ravages of predatory insects and weeds. This misrepresentation was easily countered simply by declaring that such critics had not read her book: they had not read her page-twelve disclaimer that she was not contending that chemical pesticides must never be used; nor had they read her last chapter, in which she discussed viable biological alternatives to chemicals. (Her final chapter had been added to the book in anticipation of charges that she offered no good answers herself.[83])

Although Carson did not necessarily expect the general audience to read every word, she allowed those who criticized the book in public no such leeway. The theme "They did not read the book" was one of the most common ones in public appearances and letters to the editor, voiced as ardently by Rodell as by Carson herself. Carson would offer amused derision of those who dared to embark on public discussion without having read the book, as in her speech to the Women's National Press Association, during which she read a clipping from a Bethlehem, Pennsylvania, newspaper. The news item described adverse reactions to *Silent Spring* in two county farm bureaus: " 'No one in either county farm office who was talked to today had read the book, but all disapproved of it heartily.' "[84] Carson suggested a variation on the same approach to Houghton Mifflin's Anne Ford to deal with R. Milton Carleton, of the Chicago *Sun-Times*, one of Carson's principal detractors. "I have discussed with Marie the possibility that someone might write Mr. Carleton each time one of these [misstatements] appears and simply ask him to state on which page of *Silent Spring* the statement he refers to occurs. . . . [P]erhaps if he finds he is being challenged on some of his ridiculous statements, he will be a little more careful."[85]

The false "all or none" interpretation of her message was one that concerned her from as early as her 1961 letter to the *Boothbay Register*, in which she advanced the idea that tree spraying was not a matter of choosing between "trees or birds" but rather a matter of the likely loss of both—a concept discussed in the book. She was caught by surprise when the World Health Organization (which had itself published reports concerned with dangers in pesticide use) criticized her for, in effect, ignoring the devastation of insect-borne disease, especially malaria. And although she chose not to respond directly to the organization's charges publicly, she recognized anew the need to avoid the appearance of absolutism in her cause. Before the Ribicoff committee she reiterated for the senators that "a great deal of the discussion of *Silent Spring* and of the issues has . . . been placed on an all-or-none basis, which is not correct. This is not what I advocated."[86]

Carson also attempted to do a little public credibility-impugning herself on occasion. Her disapproval of the economic entanglements in pesticide supervision and policy applied to the full array of opposition spokesmen (and men they all were, with one exception.)[87] While careful not to be too strident about the fact that many of the most vocal critics had unpublicized ties to the chemical industries, her speeches typically spotlighted the propaganda ploys of her detractors. To the Women's National Press Club, she pointed out that the chemical industry was the source of much publicity supplied to the press, including print material as well as speakers: "It is clear that we are all to receive heavy doses of tranquilizing information, designed to lull the public into the sleep from which *Silent Spring* so rudely awakened it." She listed some specific examples of academic studies funded by chemical companies, observing, "Such a liaison between science and industry is a growing phenomenon. . . . It might be a less serious situation if this voice were always clearly identified, but the public assumes it is hearing the voice of science. . . . Is industry becoming a screen through which facts must be filtered, so that the hard, uncomfortable truths are kept back and only the harmless morsels allowed to filter through?"[88] More succinctly, she told the Garden Club of America, "I recommend you ask yourself—Who speaks?—And Why?"[89]

Rodell shared Carson's battles in answering challenges, but she had her own appetite for battle as well. She wrote angry letters to editors, took on her alma mater for course material she found seriously one-sided against Carson,[90] and considered filing a plagiarism suit against the National Agricultural Chemicals Association (NACA) for using passages from *Silent Spring* without attribution in its *Fact and Fancy* broadside: "I think it would be delicious fun to threaten them with a suit—if we have a case."[91] As the volume of criticism grew, Rodell wrote to Brooks proposing an ad "full of quotes" to counter the impression that "all the scientists are on the other side." She particularly wanted to quote the inelegant phrasing ("baloney") of one particular nemesis and to note the industry source of his funding: "In other words, let's kick 'em in the teeth. (I can be elegant, too.)"[92] As Carson's strength flagged and, perhaps, as she felt vindicated by the PSAC report and her reception in Congress, she relinquished management of much of the public response to Rodell.

Why a Book?

To this point, we have somewhat circled the issue of Carson's choice of the book form to pursue her mission. An author is unlikely to say explicitly why she chooses to communicate by means of a book, but some reasons can be discerned from Carson's own actions and comments. Among other things, we have the fact that Carson chose to communicate simultaneously via magazine and book, and her attitudes toward the two different forms are illuminating. While she initially conceived of the pesticide project in terms of an article that, indeed, someone other than she might write, she quickly decided it was the topic of an entire book. Thereafter, even in correspondence with the *New Yorker*, she always referred to *Silent Spring* as a book, for which the article was but serial abridgment. Why she felt that *Silent Spring* had to be a book involved considerations of format, reach, and control.

First and simplest, Carson moved from article to book form because she perceived that the scope and complexity of the subject demanded a book's length. As the quantity of material unearthed in her research grew, she came to understand her authorial process as one of synthesis. She wrote William Shawn and Paul Brooks the same letter, explaining, "It is as though all the pieces of an extremely complex jig-saw puzzle are at last falling into place. . . . I have a comforting feeling that what I shall now be able to achieve is a synthesis of widely scattered facts, that have not heretofore been considered in relation to each other."[93] Even though the readers of both book and article would benefit, Carson still considered the book the primary vehicle, despite the earlier appearance of the series. She resisted efforts to increase the number of installments in the magazine because she felt readers would not stay with the subject long enough to get to the important material at the end, something she did not worry about with the book, which would be slightly longer but in one volume.[94] Furthermore, she insisted that those quoting her take the book as their source, not the article. Part of the reason for this had to do with payment of reprint rights; but equally important to her was the certainty that the quoted material would be complete. That completeness, offered by a book and not its abridgment, was important in maintaining the book's credibility.

Above all, however, what the book had that the magazine article did not was the fifty-five pages of reference material providing crucial legitimation of her argument. The book medium conferred an authority or *probity* on its message in a way that no other medium could, not even one with the prestige of the *New Yorker*. Enhancing that was the authoritative tone used by a book author (Rodell worried that the tone of some material was "too journalistic")[95] and, beyond that, the permanence of a book. Carson told her BBC interviewer that she had elected to write a book because a scientific paper tended to be "buried."[96] Similarly, on urging Brooks to write a book about the Great Smoky Mountains, she told him, "You should not let these things get lost, as magazine articles inevitably do."[97] That permanence was at least somewhat a matter of perception, however, for Carson acknowledged publicly the constant problem of new information and discovery that could render book-borne content dated, or worse, disproved.

One advantage periodicals might seem to have over books is that nonspecialty publications by and large have a supposedly far greater reach than any book. The *New Yorker*'s circulation of 430,000 was more than four times the initial unit production (and sales) of *Silent Spring*. Carson was well aware of the much larger circulation of *Readers' Digest* and *Life* magazine, not to mention the still broader reach of radio and television; yet she chose to identify the primary exposition of her message as coming from a book. The overarching truth of media treatment of a book is that it can extend the book's reach, spreading word of publication and usually content as well. The keystone of that truth is that a book can provide access to the full reach of the several media (see chapter 5). Early in her career, Carson had turned to writing articles because one could make money more easily that way than by writing books.[98] But her success with *The Sea Around Us* and *The Edge of the Sea* had demonstrated that a book could take Carson and her message to places far beyond those even a *New Yorker* could—to meetings, conferences, interviews, and other public appearances. For an author in the media-dominated American midcentury, a book afforded access to the media in a way that a single article might not—and in a way that would enable audiences to find her message long after the magazine had left the newsstands. Carson understood that such access meant that some substantial portion of the audience might not buy or even read her book. Her sense of mission was nonetheless satisfied by the momentum that the media offered her message. Her book *did not actually have to be read* by a large portion of the audience for that mission to be accomplished.

Media access, however, required special handling, particularly with a message likely to prompt criticism. In Carson's mind and in Rodell's, the role of the *New Yorker* was not just to extend the reach of *Silent Spring* but also to introduce it with the compounding impact of the *New Yorker*'s imprimatur. The fact that the magazine would present it in more than one part would "emphasize its importance in their eyes";[99] and although the *New Yorker*'s practice of vetting all such articles was not widely known to the public, its reputation for accuracy was.

Perhaps more pivotal for Carson was the *New Yorker*'s reputation for separating editorial and advertising policy, which afforded her the essential authorial independence that is the hallmark of the book form. Each time Carson had an unhappy experience with an editor, producer, or even interviewer, it was because of loss of control over her subject matter. Her decision not to work with Diamond, for instance, was directly related to her wish to have full control over the project. Her plan not to work with *Life* magazine arose out of unpleasant dealings in which it was clear that the editors' ideas, which were not hers, would prevail.[100] Her experience with RKO concerning its Oscar-winning film based on *The Sea Around Us* had been even more dismaying; she was barred from having any say in script changes, and she was forced to sign an agreement prohibiting her from disavowing the movie in public.[101] She was wary of radio, having learned that distortions could occur in editing as well as recording; and Rodell took on a Wisconsin university radio station for broadcasting chapters of *Silent Spring* being read aloud without permission, in part because of copyright issues and in part because Carson was extremely sensitive to the possibility that a reader's voice could convey unintended interpretation.[102]

One final but pivotal consideration regarding *Silent Spring* went generally unmentioned by Carson or Rodell and has been scarcely mentioned by most analysts since then, at least publicly—that of advertising. Lear does refer to early difficulties placing a magazine article on pesticides, attributing many rejections to "fear of lost advertising revenue, [which] only heightened Carson's determination to push ahead. They represented just the sort of scientific ignorance she wanted to combat and convinced her of the public's need for information. But if placing an article was going to be difficult, she agreed with Marie that perhaps they should consider a brief book instead."[103] This is not a minor point. If Carson as a government employee might have been muted by fear of employer retaliation, she was potentially at least as much muted by magazine editors under pressure from advertisers. Thus it was no accident that the one magazine known for its separation of editorial and business matters was the one that jumped to serialize *Silent Spring*. But more significantly, a book was the one medium where advertising revenue had nothing to do with distribution, since it carried no advertising. Carson's care to keep the book from being identified with any commercial interests illustrates how essential she deemed such independence.

As Carson told Rodell, "My proper field is books, where an author can call his soul his own."[104] She made the statement in the context of the breakdown of relations with *Life* over the jet-stream article; but her observation easily represents Carson's major reasons for making *Silent Spring* a book. In sum, the book form ensured that her arguments could be organized and taken as a whole, in proper relation to one another. It could guarantee that her documentation could appear of a piece, to establish the probity of her work. It offered incomparable access to the full reach of the American media system. And it afforded her control over interpretation and expression with as much protection from external distortion, contamination, or even censorship as possible.

CHAPTER 3

Editors and Publishers: Dealing with a "Super-Ruckus"

Silent Spring's daunting technical details and dark theme might have put off any publisher, let alone a magazine with the *New Yorker*'s reputation for light, sophisticated wit. For an august book publisher like Houghton Mifflin, the subject matter might have been less problematic, but it certainly did not suggest itself as a potential best-seller. For both publishers, the possibility of a controversy, although desirable for publicity, entailed another kind of risk if the publicity turned negative enough to threaten the book's viability. Yet both publishers did not hesitate to publish the work. Why did they make that decision? And what difference did it make for either of them that it would appear in both book and magazine form?

Although the message enclosed within a magazine or book is the author's, it is the publisher who has the most to do with how that message reaches the public. Here we will begin with a sketch of the histories of the *New Yorker* and Houghton Mifflin and some background on those directly involved with *Silent Spring*. Against that backdrop, we will investigate what went into decisions to publish Carson's work, how each publisher met the challenges to her message, and what the involvement of the two respective institutions meant for the course of the public debate about *Silent Spring*. For one publisher, it was a matter of full immersion into Carson's mission and the debate sparked by the book. For the other, it was simply a matter of putting Carson's message forth in the best form possible. The role of each institution, however, had to be taken into account by the other.

The New Yorker and William Shawn

The *New Yorker* had been founded in 1925 by Harold Ross[1], who announced his concept of the magazine's mission thus:

> *The New Yorker* will be a reflection in word and pictures of metropolitan life.
> . . . It will not be what is commonly called sophisticated, in that it will assume
> a reasonable degree of enlightenment on the part of the readers. It will hate
> bunk.
>
> As compared to the newspaper, *The New Yorker* will be interpretive, rather
> than stenographic. It will print facts that it will have to go behind the scenes to
> get, but it will not deal in scandal for the sake of scandal nor sensation for the
> sake of sensation. Its integrity will be above suspicion. It hopes to be so enter-
> taining and informative as to be a necessity for the person who knows his way
> about or wants to.
>
> *The New Yorker* will be the magazine which is not edited for the old lady in
> Dubuque. . . . *The New Yorker* is a magazine avowedly published for a metro-
> politan audience and thereby will escape an influence which hampers most
> national publications. It expects a considerable national circulation, but this
> will come from persons who have a metropolitan interest.[2]

More succinctly, Ross once said, "This isn't a magazine! It's a movement!"[3]

The uniqueness of the *New Yorker*, even among other literate or "intellectual" magazines, derived from the character of its editor in chief, in
much the same way an individual publisher's personality marked the elite
book publishing houses of the era. Although histories of the magazine have
commonly dwelled on the personality of founder Harold Ross and his
indelible imprint, somewhat less attention has been paid to his successor,
William Shawn.[4] Shawn's background was (like Ross's) neither that of
urbane New Yorker nor bookish editor. Born in the Midwest and with
only two years of college at the University of Michigan, he began as a
reporter for a small New Mexico newspaper and later became an editor for
the Chicago-based *International Illustrated News* before moving to the *New
Yorker* in 1933. There he began as a reporter and rose through the ranks,
becoming editor on Ross's death in 1951.[5] His training gave him both a
journalistic desire to bring important issues to public attention and an
intense editorial style that had him working closely with authors, thoroughly involved in cutting and revising. Paradoxically, those who knew
Shawn well noted a particular squeamishness about disease and an "aversion to the clinical,"[6] which may have made him more sensitive to Carson's
message but perhaps uncomfortable about it at the same time.

Shawn is often described as merely having carried on Ross's traditions and customs, but that description is accurate only to a point. Ross had established three traditions that Shawn carried on, though with some modifications pertinent to the serialization of *Silent Spring*. First, Ross's choice of an elite audience informed the now well-known high-culture tone of the magazine; but the working definition of that elite had several facets. Ross had targeted an urban and urbane readership; but despite the magazine's title and the local focus of some of the material (such as arts and entertainment), he had from the outset sustained the idea of a national readership. In fact—his exclusion of Dubuque notwithstanding—approximately three-quarters of the magazine's circulation eventually came from beyond the New York metropolitan area.[7] Ross had limited circulation to about 400,000, however, thus exclusivity was part of the cachet he sought and which Shawn preserved, keeping the circulation between 400,000 and a high of 500,000 in 1986.[8] By the early 1960s, the *New Yorker* was thus well established as a national organ with a limited but influential and highly desirable readership—especially from an advertiser's point of view.

Second, Ross's brand of journalism had brought an adamant respect for fact, and his goal for the magazine was a "technically flawless" magazine every week.[9] He developed a formidable system of fact-checking that became legendary in media circles. Although the magazine was probably most appealing to the larger readership for its fiction and humor, its nonfiction articles—biography, essays, profiles—established the *New Yorker* as a bastion of well-researched and serious writing. Shawn continued the tradition of unbending devotion to fact and if anything, enhanced it, occasionally to the consternation of those preferring a lighter tone and finding Shawn's magazine "grim."[10] Like Ross, Shawn had a respect for fact that meant continuing reliance on troops of checkers; but Shawn felt he had a different attitude toward facts. In the beginning, Shawn had shared Ross's belief that facts were "an end in themselves; they were self justifying." However, Shawn came to believe that nonfiction could and perhaps should be interpretive, even while concerned with fact, explaining, "It was only in later years that I realized that facts in themselves might be meaningless or worthless, or might need defending."[11]

Similarly, while Ross was generally wary of advocacy, Shawn took on difficult issues, more so later in his tenure but beginning while associate editor and later managing nonfiction editor under Ross. In 1946 Shawn had persuaded Ross to devote all of one issue to an unabridged republication of John Hersey's *Hiroshima*. That issue came to be considered a landmark for the *New Yorker*, and frequent parallels would later be drawn between Hersey's account of the horrors of nuclear war and Carson's alarm concerning toxic chemicals. Although Shawn never made his editorial philosophy explicit, he seemed to have radar for what could be a significant social issue, ripe for debate, in which the *New Yorker*'s role might well be to "defend" the facts as they came to Shawn's attention. On at least one occasion, however, he did set forth his priorities as publisher-editor. He wrote to the author of an article on the United Nations (which he eventually did not run), "I want you to know that I have the interests of your article at heart, but I also have to think of the interests of the magazine and the country."[12]

Insofar as it may be discerned, Shawn's sense of mission probably did not follow from a personal involvement with the actual subject of pesticide abuse, particularly given his discomfort with "the clinical." But by the early sixties Shawn was firmly committed to presenting subjects that would challenge the public to expand both knowledge and understanding, and he counted on that reputation to sustain the *New Yorker*'s status among the media.[13] He was undoubtedly gratified by comments like the *Holyoke (Mass.) Transcript-Telegram*'s in 1963, which called his *New Yorker* a "launching pad for ideas that jar the middle class complacency that swathes us all. It was in that magazine that Rachel Carson's *Silent Spring* first appeared."[14]

Most significantly, however, the absolute separation of editorial policy from advertising and marketing under both Ross and Shawn was a defining characteristic of the magazine. Ross had insisted on that separation from the outset and made it clear that, in a pinch, the editorial decision would always take precedence over business considerations.[15] If anything, "the business was supposed to protect us! That's what they were there for."[16] This legendary partition of commerce and content was made possible by the magazine's enormous profitability, which Shawn was able to preserve well into his tenure. By 1962, Shawn's tenth year as editor, the *New Yorker*'s circulation was limited to approximately 435,000, but it set a record among periodicals, achieving nearly $2 million in profits.[17] This profitability derived in part from the mathematical fact that it carried more advertising than any other magazine of the era. More important, however, it was a most desirable advertising medium for businesses wishing to reach an influential elite. As one marketing consultant of the era observed, "It was a club you wanted to belong to if you had a prestige product."[18]

Overall, the magazine under Shawn was in the unique and enviable position of being able to turn down advertisements that either offended the editor or that simply made the magazine too thick for Shawn's aesthetic sensibilities. Significantly for controversial pieces such as *Silent Spring*, the editor could easily ignore any displeased advertiser pressing for editorial changes or threatening to withhold advertising. Indeed, the magazine took pains to alert advertisers when some content might be troublesome for them, in case they wished to withdraw their ads from a given issue.[19]

This insulation from advertiser pressure permitted Shawn to experiment and to take risks. Excerpting or serializing an established best-seller was not uncommon, but excerpting or serializing a book before publication meant working without certainty of success. The *New Yorker*, however, had developed a cadre of well-known, reliably successful authors and a practice of publishing excerpts from their forthcoming books. Carson was already a member of that group, having had both *The Sea Around Us* and *The Edge of the Sea* excerpted in the magazine. But publishing articles of great length—or even turning an entire issue over to them as with "Hiroshima"—risked intimidating or overwhelming the readership. Many criticized Shawn for the increasing average length of favored articles, and advertisers were particularly bothered by this trend. Ultimately, the only differences between the "Hiroshima" issue and a book (other than the soft cover) were the presence of advertisements and the customary periodical front matter; and several issues under Shawn approached bookish proportions. Indeed, some have argued that the *New Yorker* of that era had more in common with books and book culture than with other magazines, because of its long sections of text with no illustrations and no jumps and because of Shawn's editorial style and outlook.[20] If the magazine acquired added stature from having borrowed some of the prestige of book publishing, its accommodative relationship with Houghton Mifflin undoubtedly benefited.

Houghton Mifflin: Paul Brooks, Lovell Thompson, Anne Ford

The history of most traditional American book-publishing houses is commonly cast in terms of tension between publishing as a business and publishing as a profession. As one of the oldest and most respected of family publishing houses, Houghton Mifflin faced that tension throughout its history, with its public face usually showing the professional side.[21] Houghton Mifflin's corporate history begins with one of America's earliest publishers, Ticknor and Fields, a Boston company that formed a partnership in the mid-1850s with Henry Houghton, owner of a printing plant known as the Riverside Press.[22] The firm saw various shifts in partnership, including an era when the location of management was split between New York, where it was Hurd and Houghton, and Boston, where it was known as H. O. Houghton and Company.[23] Through this period, other firms were acquired, including the magazine *Atlantic Monthly*. In 1880 the name became Houghton, Mifflin and Company (reflecting the elevation to partner of George Mifflin, who had joined the firm much earlier); and in 1908 the firm was finally incorporated. Boston remained its headquarters thereafter, removing it somewhat from the cosmopolitan New York publishing culture and allying it with the scholarly Brahmin atmosphere of literary New England.

By the mid-twentieth century, Houghton Mifflin had a long-established reputation for publishing "quality" books—fine general literature and especially texts and educational materials. The corporate history offers a long list of authors beginning with Ticknor and Fields's earliest: Henry Wadsworth Longfellow, Oliver Wendell Holmes, John Greenleaf Whittier, Ralph Waldo Emerson, Harriet Beecher Stowe, Nathaniel Hawthorne, Henry David Thoreau, and Mark Twain. The company was also the authorized American publisher of many English authors, including Alfred, Lord Tennyson, and Charles Dickens.[24] Over time, the list grew to include Esther Forbes, Henry Adams, Samuel Eliot Morison, Willa Cather, John Dos Passos, and Amy Lowell. The publication of Winston Churchill's six-volume history of World War II was a notable moment in the company's history, oddly balanced in one history with Hitler's *Mein Kampf.* In the postwar era, Houghton Mifflin's education department grew along with the baby boom, while the tradition of high-quality trade books continued. The pride of the company in its literary heritage was expressed in a 1999 corporate history in terms unmistakably reflecting a sense of mission involving a contribution to public thought: It noted that the humble eighteenth-century storefront beginnings "gave little indication of the great influence that the ideas contained in our publications would have on American society during the next 150 years."[25]

A family feeling within Houghton Mifflin was also a hallmark of the house, which a 1964 trade magazine profile found "astonishing in a company large enough to have 1,448 employees and a backlist, which even after a recent winnowing of less active titles still totals over 2,300."[26] That sense of family and tradition had a fiduciary counterpart, however, in that ownership was closely held. As John Tebbel noted, "The house was a corporation, . . . but the stock was tightly held by a few people, most of them in the company."[27] In the era of *Silent Spring* the company was still resisting the trend of "going public," and Houghton Mifflin was to be among the last of the privately held family publishing houses to be opened to public ownership, holding out until 1967. Even after 1967 Houghton Mifflin remained remarkable for its independent status until June 2001, when European conglomerate Vivendi Universal purchased the venerable house.[28]

Paralleling the succeeding generations of family owner-publishers were epochs of near-dynastic sequences of publishers and editors in chief, who often then became vice president, president, or chairman of the board, for many of these executives spent their entire careers in the same company. Reflecting that pattern, the executives most involved with *Silent Spring*— Paul Brooks, editor in chief; Lovell Thompson, executive vice president for trade books; and their boss, William Spaulding, chairman of the board—had already spent much of their professional lives with Houghton Mifflin.

Other than literary agent Marie Rodell (and of course, Carson herself), no one was more involved and invested in the career of *Silent Spring* than Paul Brooks. Much of the story of *Silent Spring*'s career at Houghton Mifflin is the story of Brooks's devotion to author, book, and message, all three of which became pivotal in his own life.[29] Educated at Harvard, Brooks had started with Houghton Mifflin in 1931 as a reader, making his way to editor in chief by the 1960s, and retiring as vice president in 1969 to make writing his primary occupation. In his memoir of his time at Houghton Mifflin, he described the era as "a time when book publishing was as much a profession as a business, with the personal relation between writer and editor at its core."[30] His own relationships with his authors were close and even social, though this was not unusual among authors and editors in American publishing then, in part because they often ran in the same social circles. He had presided over Rachel Carson's arrival at Houghton Mifflin in 1953, but he had known her and her work since 1950; and over time their social relationship grew to extraordinary depth. A shared love of nature and mutual respect for their professionalism forged a bond that was to extend even beyond her death in 1964, when Brooks and his wife became foster parents for her orphaned grandnephew.[31]

An author himself, Brooks later produced several nature books, as well as a book on Carson's life, and his work earned him some prestigious awards.[32] He was thereby more understanding of the author's tasks and

Paul Brooks, editor in chief, Houghton Mifflin. With thanks to the Brooks family.

trials, particularly in writing about technical matters. For him, the risk in publishing had to do primarily with the outcome of authorial and editorial creativity. He wrote in an *Atlantic* article, "A book contract is the record of an act of faith. It is necessarily so from both the author's and the publisher's point of view. . . . It involves a continuously creative process with all the uncertainty and risk that implies. It can never guarantee results."[33] Beyond their relationship as editor and writer, Brooks and Carson shared a love of nature that was almost as defining in Brooks's life as it was in Carson's. Devoted to conservation of wildlife and natural habitats, he had become a director of the Massachusetts Audubon Society in 1943. In addition, he was a member of the Nature Conservancy and a director of the Massachusetts Trustees of Reservations (and would later become member of the Sierra Club national Board of Directors). When Carson became a Houghton Mifflin author with *Edge of the Sea* in the mid-1950s, the company was undertaking to expand its nature list, very likely under Brooks's motivation and guidance. Brooks was thus both qualified and inclined to edit and publish works on the natural world; but more significantly for *Silent Spring*, he was already committed to its philosophical underpinnings.

Because of his Ivy League background and his interest in nature, Brooks had a built-in network of like-minded associates, notably in the Audubon Society but also in many other quarters. Interior Secretary Stewart Udall knew of Brooks and his interests, and while the President's Science Advisory Committee's (PSAC) report was still under wraps in early May 1963, Brooks received an informal letter from a contact in Udall's office addressed and signed with first names, alerting him somewhat confidentially to the imminence and probable leaning of the PSAC report. Otherwise, thanks to Brooks's contacts among conservationists and nature lovers as well as in academe and government, he had access to many channels of communication from the most local to national and even international levels.

Lovell Thompson, executive vice president for trade books, was far less personally engaged in causes such as conservation or nature, although he found himself increasingly enthusiastic about *Silent Spring* as the fray gathered energy. Harvard-educated like Brooks, he was literate and business-savvy as well. He told a trade magazine, "We publish about 100 trade titles a year, . . . and we maintain our general size and stature with fewer titles than our competitors. This means that our sales per title tend to be larger than average," which, according to Thompson, meant that an editor had to know "which pool to fish. . . . He knows he has to be in an area where sales exist" in order to have a chance at finding a best-seller.[34] But the attention to business never eclipsed Thompson's commitment to Houghton Mifflin's "higher" calling. Brooks wrote of Thompson, "For a quarter of a century he was a leading figure in shaping the character of our enterprise. Concerned more with good books than with quick and easy profits, he proved that excellence can indeed be profitable."[35] Thompson's ability to straddle the two roles of protector of Houghton Mifflin's reputation and manager of its commercial fortunes was determining in how the firm handled *Silent Spring*; and an openness to controversy was key to his success. In 1944, already head of the trade department, Thompson had written, "There is nothing either businesslike or financial about publishing. . . . No one ever went into publishing to make money; we are in it because we like the written word and the semi-public commotion it can always kick up."[36]

Finally, the person most directly responsible for handling any "semi-public commotion" arising around a Houghton Mifflin book was Anne Ford, head of publicity. Although little is recorded about Ford, a salient trait was her own journalistic orientation. At the time, promotion for publishing houses was often carried out by women with journalistic backgrounds, who found perhaps more professional opportunities in book publishing than in print or broadcast news media. According to Lear, Ford had been a journalist before working in book publishing, taking her position at Houghton Mifflin in 1953, just before publication of *Edge of the Sea*.[37] In correspondence, she sometimes made a point of referring to herself as a former journalist, and she was often compelled to remind Brooks and Thompson of the realities of dealing with the media. Though there are no surviving indications that Ford had any specific personal interest in nature or conservation, she was related by marriage to a prominent nature writer who was also a friend of Carson.[38] Her correspondence on behalf of Houghton Mifflin reflects an enthusiastic, professional woman, loyal and highly partisan on behalf of the company's authors. Although on occasion she and Rodell did not entirely agree on strategy and tactics, she worked diligently to maintain a good relationship with Rodell, especially as both women became protectors of the increasingly frail Carson. Ford was often in the position of mediating between Carson and the press (or between Rodell and Thompson), particularly in the matter of how much to try to control the news and how much to respond to the news. Inevitably, Ford's position sometimes put her in the center of the controversy.

Publishing, Promoting, and Defending *Silent Spring*: The New Yorker

Carson's original conception of the size, shape, and depth of her work on pesticides evolved largely under the guidance of William Shawn. His involvement in the early stages of framing and conceptualization seems to have been considerable, and his editorial involvement with the actual writing of the piece may have exceeded Brooks's. It was partially because of Shawn's contribution to the work that Carson became sole author, since for Shawn and the magazine, "co-authorship on the *New Yorker* [was] unthinkable."[39] As Carson and Shawn discussed the project in 1958, he came to envision it as a "long, two-part piece,"[40] and its dimension and shape were altered. "Mr. Shawn's ideas completely change the picture, for the *New Yorker* piece we visualized could, with only slight amplification, stand for the complete book. . . . If I do 20,000 to 30,000 words for the *New Yorker*, ranging over the entire field of subject matter, this will certainly be the heart of the book"— a book for which she would ultimately be exclusive author.[41] Moreover, the *New Yorker*'s publication schedule, not Houghton Mifflin's, drove the timing of publication. As Carson wrote to Rodell in 1958, "While I hope for completion by the end of June, I cannot guarantee it; nor do I feel that a book publication deadline is the controlling factor."[42]

As work on the project stretched into years, Shawn and Carson conferred frequently, and she submitted parts in draft through 1961. But it was not until January of 1962 that a full manuscript was sent to Shawn's office, in the form of fifteen chapters sent via her agent, with a copy also sent to Brooks at Houghton Mifflin. Shawn called her at nine on a Monday evening, and to her joy, his response was more than positive: "A brilliant achievement . . . you have made it literature full of beauty and loveliness and depth of feeling."[43] With the manuscript in hand, Shawn began to see its length expanding well beyond his original estimate. At one point he envisioned it as a six-part series, and Carson was probably only one of several prevailing on him to limit the series to three parts.[44] With publication now scheduled for early summer 1962, Shawn applied his special genius in working with Carson to make the technical aspects of her exposition more accessible and more compelling for the readers, by combining "diligent reporting with an appropriate literary form."[45] Shawn's literary form entailed organizational changes that postponed scientific information and Carson's disclaimer until after more dramatic statements and description. The effect was rather more journalistic than scholarly, even though the quantity of technical material meant that no one could mistake the series for sensational reporting.

It was not the *New Yorker*'s custom to pursue extensive publicity for an article or series. The magazine did maintain a list of several thousand recipients of complimentary issues, sent almost entirely to advertisers as a demonstration of continuing high quality, but recipients also included influential "VIPs" in New York and Washington.[46] Presumably, the issues with Carson's series were circulated to that "comp list," and the usual alert about content of special note was provided to editors on important newspapers, including the *New York Times*, which published an editorial praising the series more than a week before it appeared on newsstands.[47]

Although once an article appeared, the *New Yorker* declined, on general principles, to participate in any further discussion of it in any forum, the magazine did take considerable pride in having provoked comment in the mainstream media. Press commentary and letters to other editors were always included in the *New Yorker*'s promotional newsletter, "The Quoter."[48] Letters to the *New Yorker* editor were not published (Shawn's name as editor did not even appear on the masthead), nor did the *New Yorker* customarily respond to commentary or editorials in other media. The number of letters written in response to the *Silent Spring* series broke *New Yorker* records; but no mention of their existence, let alone their content, was ever made in the magazine. Many readers' letters were privately answered, however, particularly if they involved a specific inquiry, such as queries about the identity of the baby-food manufacturer mentioned in *Silent Spring* that tested its products for pesticide contamination (these letters were referred to Carson; the firm was Beech-Nut). Requests for reprints drew the stock response that a book was forthcoming from Houghton Mifflin and that therefore, for copyright reasons, the *New Yorker* would be providing no reprints. The tone of that response was at worst neutral and at best an encouragement to buy the book.

The *New Yorker* thus took part in generating the public debate but declined to participate in it further—a fact that distinguished it from all other participants in the career of *Silent Spring* but one that does not necessarily negate the significance of the magazine's implicit support for the work. With a less controversial piece, the work of the *New Yorker* would have stopped at editorial efforts, but Carson's opponents did present the magazine with a challenge. On June 20, at the approximate time of the second article's appearance, the *New Yorker*'s legal counsel, Milton Greenstein, received a telephone call from Louis McLean in Chicago, who identified himself as "an attorney representing two chemical companies." Although McLean indicated that "he was not talking for publication and that there was no question of litigation," he listed a number of alleged inaccuracies in the first installment of the series. He urged the *New Yorker* to consult certain named authorities who would dispute Carson's findings. And although his demeanor during the call was reported as "friendly throughout," he indicated that "in the interests of fairness, it would be better not to run the third part of the article." Greenstein summarized for Shawn, "There was no doubt in my mind . . . that McLean was attempting an intimidation and hoping to effect what changes and modifications he could, to reduce damage to the chemical industry as a whole." Greenstein stood firm. "I told him I didn't think there was the slightest chance that there was any information available that could persuade us not to complete the series,"[49] he reported, having said to McLean, "Everything in those articles has been checked and is true. Go ahead and sue."[50] (McLean then turned his attention to Houghton Mifflin.) The controversy's only direct effect on the magazine was thus dispatched with minimal notice, and the magazine's reputation for integrity remained intact, even out of the public eye.

If any advertisers objected to the content of the three issues, no record of it could be found. Any such objection would most likely have been directed to the sales department, from whom Shawn did not want to hear and did not have to hear, given the charter of the magazine. Also, although the *New Yorker* drew advertising dollars from almost all sectors, its greatest draw was among prestige retailers of consumer products like clothing, jewelry, and automobiles—not industries potentially affected by questions about pesticides. However, a review of *New Yorker* issues before, during, and after the three installments of *Silent Spring* shows that three chemical producers routinely placed full-page ads in the *New Yorker*, not always on a weekly basis but often more than once a month: DuPont (advertising nylon stockings), Cyanamid (advertising a variety of chemicals, such as those for water treatment), and Esso/Humble. All three apparently continued to do so without noticeable interruption throughout June and July of 1962. Indeed, Esso ran a special two-page spread in the issue carrying Carson's final installment, focusing on the survival of whooping cranes "with an assist from oilmen."[51]

The *New Yorker*'s role in the publication of *Silent Spring*, according to its standing journalistic practice, was greatest in the conceptualization and creation of *Silent Spring* as a story, largely limiting general publicity of the story to the simple, literal fact that the series appeared in its pages. Its job as a magazine was essentially to record and circulate the information—Carson's facts—underwritten solely by the magazine's reputation for factual accuracy. Although she worked many references into the articles' text, formal documentation of her sources was not included; thus "defense" of her facts hinged on the solidity of her arguments and the magazine's reputation. That defended truth was, however, more than mere communication of organized information. As Shawn told Carson, "After all, there are some things one doesn't have to be objective and unbiased about—one doesn't condone murder!"[52] That philosophy and his editorial role took Shawn well past mere gatekeeping or even setting an agenda item for public debate, into the realm of outright advocacy. Shawn welcomed the role, although it had not been the customary one for the *New Yorker*: "We don't usually think of the *New Yorker* as changing the world, but this is one time it might."[53]

Publishing, Promoting, and Defending *Silent Spring:* Houghton Mifflin

Compared with the *New Yorker*, Houghton Mifflin, as a book publisher, could expect to have far wider and deeper functions with respect to the author and her message—not only editorial and production efforts but also substantial promotional and public-relations functions before, during, and after publication. A single book is, after all, a far greater part of a year's effort in a publishing house than a single article—or even a three-installment series—is for a weekly magazine. On the heels of the *New Yorker* series, however, Brooks and Houghton Mifflin were confronted with rising controversy over their forthcoming book, one that demanded increasingly more effort as time went on. Their task became not only to do what was necessary to ensure exposure and sales but even more to preserve the public viability of the book and the credibility of its message, its author, and—by association and extension—Houghton Mifflin itself.

In the years during preparation of *Silent Spring*, Brooks had sole responsibility for the Carson project. He had proposed it to Houghton Mifflin's Executive Committee in April of 1958, indicating its timeliness by referring to the Long Island aerial-spraying lawsuit and identifying other experts concerned with the problem. To downplay any possible controversy, Brooks told them, "This is not to be simply an attack on present methods of nature control, but rather an interpretation of the entire biological community, showing that there are other and more successful ways of controlling nature in the interests of man." Anticipating worry about the riskiness of publishing too technical a book, he wrote, "Though the subject may at first seem rather specialized, I believe that there is a fairly large market waiting for this book," especially given the "clear parallel to the problem of nuclear fallout."[54]

Later as public rumblings about environmental problems continued, he drew the same parallel for Carson, predicting receptivity to her project: "In a sense, all this publicity about [radioactive] fallout gives you a head start in awakening people to the dangers of chemicals."[55] He took care, however, to avoid too slavish a link to news that might or might not interest the general public. For the title of the book, Carson had proposed using the title of the *Saturday Review* reprint of Justice William O. Douglas's dissent to the Supreme Court ruling in the Long Island lawsuit; but Brooks was unenthusiastic. "I don't think that 'Dissent in Favor of Man' is quite it. Speaking of titles, 'Silent Spring' might be a good one for Chapter B."[56]

Like Shawn, Thompson and Brooks had reservations about having anyone other than Rachel Carson be the author of the work. Brooks later noted, "All the way along I was wanting *her* to do it. . . . You always want a book like this done by one person. . . . I'm very skeptical of collaborations" on both editorial and aesthetic grounds.[57] Thompson feared loss of control over the tone of the work: "It seems to me this thing begins to grow dangerously synthetic. . . . How much of it will Rachel really write? . . . We need someone . . . who feels passionately and who . . . has time to be the fiery heart of this book."[58]

As editor, Brooks was charged with understanding the potential reading audience, and like Rodell, he was watchful concerning the accessibility and tone of the book. The chapter outline Carson sent Brooks in December 1959 gave him an idea of the shape and breadth of the book, as well as its probable appeal: "The immediate human application of all this is what's going to sell the book," he wrote her.[59] His own worries about the technical nature of the subject surfaced from time to time as work progressed, however, and his editorial comments about some of her chapters indicated his shared concern with Rodell that certain material might be too rough going for the intended lay reader. When he contacted his friends and naturalist allies to spread the word about the book to come, he acknowledged to them that "even with Rachel's name this isn't necessarily going to be an easy book to sell since the subject itself is rather forbidding."[60] As he put it years later: "Who wants to read a book about pesticides?"[61]

Illustrations might facilitate things, but care had to be taken. Thompson had proposed one illustrator whose work was occasionally humorous, but Brooks objected: "I doubt whether there are many places for humor in *Silent Spring*."[62] When Brooks wrote to the eventual illustrators (Lois and Louis Darling), he indicated that the artwork should be limited: "I think there is a real danger of having this book over-illustrated, and conceivably making it look like a children's book."[63]

As Carson submitted material to Brooks, his enthusiasm grew perceptibly and with it his personal commitment to getting the word out. "I was her first convert," he declared.[64] Early on, he encouraged her by telling her that "one thing you can be sure of: the world is waiting for it."[65] A year later, in 1959, he had begun to perceive that considerable controversy was likely, and he welcomed it: "We can certainly gird ourselves for a super-ruckus when *Man Against the Earth* [the working title] appears. I look forward to it."[66] Like Carson, he knew that surviving the "super-ruckus" would depend enormously on the work's credibility. "The growth of the project is the principal reason that it has taken so much longer than any of us expected. . . . [It] keeps on being one book with a larger and larger background of fact behind it. When the iceberg is finally launched, we will try to make people realize how much lies beneath the surface."[67]

In February of 1962 as the manuscript was in final preparation, he wrote to Carson, "The more time I spend on this manuscript, the more I feel that we *must* get people to read it; not just those who already understand the problem but the thousands and thousands of intelligent persons who would care deeply if they only knew."[68] He also made his commitment to the cause known to his contacts around the country, writing to Audubon national president Carl Buchheister, an important ally in the ensuing debate: "I think we shall have to work up something of a crusade if we are going to put it across. After reading the manuscript I feel the subject is so important that I am willing to go to great lengths to see that the book is read by the public at large, not just by the specialists."[69] In enlarging the scope of the book's potential audience, Brooks was—perhaps unwittingly at this point—allowing for the possibility of a "super-ruckus" much greater than Thompson's genteel "semi-public commotion."

While Carson had worked closely with Shawn on editing the articles, her interactions with Brooks on editorial matters were somewhat less intense and were sometimes carried out through Rodell. Brooks felt able to rely on Carson's instincts as a writer and on the thoroughness and scrupulousness of her research methods, including her verification practices. Despite organizational differences, deviations in the actual wording between the book version and the magazine version would have opened up vulnerability to conflicting interpretations, which so controversial a message could ill afford. Thus Brooks was, in effect, relying indirectly on Shawn's editorial skill and the *New Yorker*'s fact-checking as well. But where needed, and where Carson herself requested, Brooks saw that her text was further vetted by appropriate readers.

In February 1962 Brooks wrote Carson, "Well, your job is almost done;

then it will be up to us";[70] and by the time the *New Yorker* articles appeared in June 1962, Brooks's editorial work was almost at an end, save for supervision of final production details like size of print run and other mechanical decisions. But Brooks's supervisory work regarding all aspects of Houghton Mifflin's care of the book was just beginning as the anticipated "super-ruckus" bubbled to life in the summer of 1962.

Brooks was not the only person within Houghton Mifflin to perceive the likelihood of controversy, nor was he the only one to be caught up in the cause of *Silent Spring*. Thompson had some inkling of a splash, which he welcomed in the interests of promoting the book. In a handwritten note concerning his dislike of the working title, "Man Against the Earth," he wrote to Brooks, "Its [*sic*] an ill poison that doesn't give some publisher a good book and what do we call it WHO ARE YOU KILLING or LETS [*sic*] DIE IN BED or OF MONSTERS AND MOSQUITOES, . . . but it *does* sound good in fact a landmark* *if you'll pardon the expression."[71] Words like "cause," "crusade," and "our side" crept into interoffice correspondence as the spring progressed, and the atmosphere intensified until well past publication. A member of the marketing department felt compelled to write Carson directly in May of 1962: "All who have read [*Silent Spring*] at Houghton Mifflin seem to feel a spontaneous and unusual feeling of the importance of the book. Certainly, in the capacity in which I work, I shall make special efforts to see that your message will reach the widest possible audience."[72] That simple declaration reflected the sense of mission shared by all involved at Houghton Mifflin.

Brooks's own sense of mission worked to focus his concern on the most local of audiences. He wrote Carson, "This is not an easy book to tell people about. We are going to have to work up something of a crusade—on a local level—if we are to reach a really wide audience. For example, our local paper in Lincoln and Sudbury, which printed the letter on mosquito control that you wrote me to read at our Town Meeting, will undoubtedly run something about the book when the time comes. I should like to see this multiplied by a few hundred thousand."[73] His preference for attention and eventually action at the local level mirrored Carson's own.

Houghton Mifflin had an all-out promotional plan in the works that also targeted the local as well as regional and national levels. Once the manuscript was in hand and the true dimensions of the likely controversy began to be evident in early spring 1962, a full campaign was launched. Prepublication notification went to the trade journals as soon as possible.

Publishers' Weekly's first notice was an informal paragraph in mid-April, but formal notice there and in *Kirkus* and *Library Journal* came in the summer (rather late considering the June appearance of the *New Yorker* series).[74] The usual lists of book editors and reviewers were expanded, particularly as clippings began to come in from regional and local papers when the first *New Yorker* installment appeared.

Advertisements were placed in major and regional newspapers, magazines, and retail trade publications. The "travelers" (field representatives who visited bookstores, libraries, academic institutions, and other likely organizations) had been alerted that the book would generate attention and controversy. In some cases, particularly in the agricultural Midwest, field managers were feeding information back to Houghton Mifflin about local response to news reports concerning pesticides, fallout, or related environmental issues; they predicted great interest in the book. Because of Houghton Mifflin's growing list of nature books, many travelers already had relationships with local Audubon, garden, conservation, and nature club members, expanding well beyond Brooks's own national network of personal contacts. Houghton Mifflin was unusual among educational text publishers in that it made sure travelers to academic institutions were also aware of Houghton Mifflin trade publications, often using complimentary copies of best-sellers as inducements to purchase a textbook series. *Silent Spring* was to become one of the most sought-after such "comps" for academic buyers.[75]

Sales of secondary rights were always cause for celebration but posed their own kinds of difficulties with respect to *Silent Spring*. Negotiations with Consumers Union (CU) for a special *Consumer Reports* edition had to work around possible conflicts with paperback-rights negotiations and the need to preserve the integrity of the text. CU was pressed to agree to limit distribution to members only; and in an unusual departure from its policy requiring in-house editing, CU accepted editing entirely under Carson's supervision. The close timing between the *New Yorker* articles and the book publication date also made other negotiations difficult. The Book-of-the-Month Club held its announcement of *Silent Spring* as its October selection until a few days after the first *New Yorker* installment had been on the newsstands, conceivably so that the first responses could be gauged. Other reprint and syndication requests were largely denied, with exceptions made for a handful of magazines and newspapers, all timed carefully to avoid conflict with hardcover Christmas sales and future paperback sales.

The close timing between magazine and book publication also complicated the customary provision of preview copies to important reviewers, and in April 1962 Rodell and Ford crossed swords on the point. While Rodell saw great advantage in getting proofs to such supporters as Justice Douglas as soon as feasible, Ford worried about tipping off "our enemies" too soon, not to mention putting off potential readers. She wrote Rodell, "[Sending out review copies] should come after the *New Yorker*, indeed after Labor Day, when people are home from holidays and ready for a Cause and are reading home town papers again. But we don't want to frighten the average reader and let him think we have a propaganda book."[76]

For reasons unrelated to the *New Yorker*'s plans, release of the proofs turned out to be "a very tricky moment," as Brooks later described.[77] As inquiries and requests for review copies began to come in from possibly hostile reviewers, potential legal issues loomed. The possibility of a lawsuit had occurred to everyone, and certain passages and the bibliography were still being checked even as some proofs were sent out. Ford wrote anxiously to Rodell, "Would you be sure that any proofs that are sent out are clearly marked 'uncorrected proofs,' and the lawyer also told us we should write on them 'no quotation please.' This is to doubly insure us the proofs couldn't be borrowed by anyone and get into any hands that might possibly quote from them not realizing they were 'uncorrected proofs.' "[78] Anything that might undermine the integrity of that great mass of research beneath the "iceberg" threatened the entire enterprise.

Other "tricky moments" involved legal questions. To protect credibility and to ward off charges of hypocrisy, Carson and Rodell were adamant about dissociating themselves from any commercial interest or product. As a manager with a mind for business, Thompson might have been somewhat more open to possibly synergistic benefits of cooperation with other businesses, although he did agree to include a paragraph disclaiming any commercial associations in all printings after the first. He was similarly less concerned than Rodell about requests in July from DuPont for ten review copies, which had unsettled her. Rodell feared that the company was seeking ammunition for a possible suit, which she wished to fend off because a suit could threaten positive publicity or even publication. Thompson's attitude was "if there is going to be a suit we'd better know it,"[79] later responding to Rodell's growing alarm:

> The only danger in such a course is the remote possibility that a bad court might grant an injunction on inadequate evidence. That could be ruinous, but

it could happen as easily (perhaps more easily) if we were to delay sending books until they are in the stores in early September. Moreover, we would then be robbed of one line of defense which would go as follows: "You've had the books since early August. If you were going to object, that was the time." I suppose also that there might be something overlooked in the book which was damaging and consequently exposed the author to an enormous suit. Any such danger is greatly lessened by the prompt sending of advance copies.[80]

While Thompson's somewhat outrageous suggestion that Carson and all her checkers might have overlooked something may have been intended merely to twit Rodell, his appraisal was clearly that the risk of suit was low and could be handled. Eventually, Brooks diplomatically (and perhaps too optimistically) told Rodell that the firm had learned that DuPont had "refused to go into the manufacture of most of these pesticides as being too dangerous, and are working on hormonal (if that is the word) control. I am sure that their attitude toward *Silent Spring* is entirely friendly, though they have not wanted to seem holier-than-thou vis-à-vis their rival firms."[81]

In early August, Louis McLean, the lawyer who had telephoned the *New Yorker*, wrote a formal letter to William Spaulding, board chairman of Houghton Mifflin. Writing on letterhead stationery, he was now clearly identified as secretary and general counsel for Velsicol Corporation. The five-page letter noted awareness of a book intended for publication and accused it of disparaging products manufactured by a single company and of containing misstatements and inaccuracies. McLean used the *New Yorker* articles as text for detailed analysis, drawing Houghton Mifflin's attention to specific passages, defending general use of pesticides, and countering the arguments with those by other "unbiased" authorities. The expressed objective of the letter was to refer Houghton Mifflin to those experts and to suggest sternly that there could be "unfortunate consequences" if Houghton Mifflin were to publish *Silent Spring* in essentially the same form as in the magazine articles. Although McLean allowed that the decision to publish was Houghton Mifflin's, he did convey Velsicol's insistence that its own products, chlordane and heptachlor, be treated accurately and fairly.[82]

The fact that no particular "unfortunate consequences" had befallen the *New Yorker* did not keep Houghton Mifflin from reacting with concern. With Spaulding on vacation, it fell to Brooks to meet the challenge. The passages noted in the letter were compared with the manuscript, prompting a follow-up letter to McLean requesting more specifics about Velsicol's problems with the text—plausibly as much a delaying tactic as a genuine request for information. In the meantime, the entire book was referred for vetting once again to Houghton Mifflin's lawyer and to a respected toxicologist, who approved essentially all of what was written. Brooks then wrote to McLean that he had satisfied himself that all was in proper order.[83] Internally, the opinion was that "the inference made is that they are in a position to instigate an action, not that they intend to. They have not proved in any way that they have sufficient evidence to take such action. It all seems a big bluff."[84]

As with the feared action by DuPont, the most serious damage to Houghton Mifflin would probably not have been actual prevention of publication but rather delay and undermining of the book's power. At the very unlikely worst, Velsicol could only have affected final publication of passages affecting itself. But even a temporary injunction would have meant delay, causing problems with syndication and secondary rights (among other things) and endangering deals with, for example, the Book-of-the-Month Club, which named the book a premier selection available one month after publication rather than the usual three. Even an unsuccessful suit for injunction, if reported in the press, could have raised doubts about the solidity of Carson's research and arguments. But with competent vetting of the manuscript and appropriate assurances from Houghton Mifflin's lawyer, Maurice Greenbaum, that Velsicol likely would not take further action, publication proceeded.

If the DuPont and Velsicol episodes may have seemed to spark a hint of paranoia, words like "sabotage" and "enemies" were appearing from time to time in the correspondence in and out of Houghton Mifflin as the controversy grew. Memos went back and forth trading information about activities in the opposition's camp: "Unconfirmed gossip. Frank Moore, Dept. of Agriculture is working on a blast on Rachel's book."[85] "A gentleman named P. Rothberg, President of Montrose Chemical Company in California, has blasted Rachel in a speech or an article. The National Agricultural Chemical Association and the NY Chemists Association are busily working to refute the book. . . . Mr. Cowan [of the *Christian Science Monitor*] seems to feel the industry is going to play dirty and really attack Rachel."[86] Moreover, increasing numbers of requests for review copies or advance copies were indeed arriving from possibly unfriendly quarters. A memo from Anne Ford to a marketing colleague suggested caution despite her eagerness for publicity: "Do you know who these people are? I am a little wary of chemical companies on the Carson. Many of them will be having an axe to grind and might be out to sabotage the book."[87]

The challenge at this point was to balance promoting and spreading the word with preserving the book's integrity and shaping its impact—and to do so on Houghton Mifflin's own timetable. In early August as the scope of the controversy was becoming more evident, Ford wrote to Rodell that the book was "looming up more controversial than I had thought." About prepublication interviews with Carson, Ford counseled, "My advice as an experienced newspaperwoman is to have her charmingly say for the present 'no comment.' "[88] The rhythm and degree of exposure were becoming very "tricky" to manage, indeed. Ford told Brooks, "One of the things from a publicity point of view I have been worried about is the fact that Rachel's book may be so well refuted before it is on sale that it will lose a lot of the great interest we have been building up through the publicity department." Thompson responded by hand on the same memo, circling the word "refuted" and writing, "I don't think so. Rather, people could just get sick of the subject feeling that Rachel had taken care of it."[89] In September, just before the official publication date, Thompson expressed similar concerns about early newspaper syndication: "If Rachel were unknown, or if her book had not so much and so effective advanced publicity, early syndication could not help but increase the number of people who know about the book. However, we were all a little afraid that coming early in the game, syndication might enable some who would otherwise buy the book to say to themselves: I read a bit of it, and I know what it's about."[90]

As Thompson implied, the *New Yorker*'s publication had effectively supplanted syndication early in the book's public life (although the book would still be syndicated later); and the ensuing press attention threatened to take away Houghton Mifflin's control over how the book would be perceived by the public. Houghton Mifflin, particularly in the persons of Brooks, Thompson, and Ford, was internally elated at the early explosion of prepublication sales. In fact, financial success was probably guaranteed at roughly the point when the Book-of-the-Month Club announced *Silent Spring* as a premier selection. But it was the nonfinancial objectives of those at Houghton Mifflin (as well as their interest in preserving the long-term marketability of the book) that dictated close attention to what was happening to *Silent Spring* and Rachel Carson in the public eye. The credibility of both as well as that of the publishing house itself was at stake.

The "super-ruckus" had gathered considerable steam between the June magazine publication and the September 27 official publication date for the book, forcing Houghton Mifflin's advertising to shift emphasis—from announcement of an important new, "shocking" book by a well-loved author to support and defense of a much-discussed book by an expert author. Their personal zeal for the work sustained Brooks and Ford especially over the months between publication and the climactic events of the following year; and their mission expanded in the direction of Carson's own mission: not only to inform and persuade but also to mobilize the public. In notes to help Carson with her November appearance at the Women's National Press Club in Washington, Brooks wrote, "Something has been started. The important thing now is to see that it doesn't stop in talk, that some positive action be taken."[91] An expectation of local activism had occurred to Ford as early as April: "I have a feeling we should get the book to some newspaper editors or prominent club women in some of the smaller cities and towns where the offense is greatest. Editorials and town meetings would result I think."[92]

Houghton Mifflin's advertising increasingly reflected the same orientation, noting both that readers were calling for action and what forms that action was taking. Ads following the release of the PSAC report carried copy explicitly pitched to the voting reader: "If you are properly to assess legislation now being prepared to control the use of pesticides, you owe it to yourself to read [*Silent Spring*]."[93] The firm's largest and latest promotional publication (not appearing until spring of 1963) concluded with a section titled "What Next?" that detailed actions being taken "in response to the demands of an awakened public," including legislative initiatives and local bans on aerial spraying.[94]

By mid-November Houghton Mifflin decided the time had come to respond directly to the criticisms now being voiced across the media system. Brooks and Ford worked with the advertising department to create an answering ad laden with supporting quotes and information intended for the general press. Brooks sent proofs of the ad to "our major list of editorial writers, literary editors, 500 or so book stores . . . and to about 60 conservation organizations. I've written a personal letter to each of the latter (of course in a number of cases I know the man in charge). . . . I'm sure that some of the Audubon Societies, for example, want to have exactly this material to support their position."[95] Brooks's networking strategy certainly worked to support the credibility of the publishing house's major product for the year. But the motivation for his letter writing was as at least as much personal as professional, and occasionally his enthusiasm seemed to blur the lines between his personal commitment to a cause and his professional role at Houghton Mifflin.

Like Carson, Brooks sought to be careful lest any appearance of financial interest undermine the strength of the argument. In February of 1963, he wrote to a University of Rochester physician of his acquaintance, asking that the doctor—rather than Brooks himself—respond to a "shockingly biased" article in *Today's Health* magazine. Brooks felt that "someone in authority should write a letter to the editor, setting the record straight. I can think of no one better equipped to do this than yourself. . . . Obviously anything that I wrote myself would be suspect as coming from someone financially interested in the success of the book."[96] In another exchange between Brooks and Ford, about the *New York Herald-Tribune*, Ford cautioned Brooks, "I think there will be plenty of letters written in without the publisher or author having to write too many of them."[97] But sometimes Brooks had to rein Ford in on the same issue. When Ford wanted to demand that a *Boston Globe* editor publish letters of rebuttal to the chemical companies' campaign, Brooks suggested they find some scientists to write the letters rather than anyone at Houghton Mifflin.[98]

Ford was as energetic a letter writer as Brooks, but her contacts were primarily those in the media. She made it her business to contact pivotal commentators if she caught wind of trouble, referring them to scientists or experts to second Carson's findings. Eager to keep the weight of the controversy solidly on "our side," she had sought permission from the *New Yorker* to tell the rest of the media how strong and positive the public response to the articles had been. Her letter to Robert Cowan of the *Christian Science Monitor* (who had raised the question of whether Carson might have told only one side of the story) illustrates her frequent practice of referring one part of the press to another. She mentioned the letters to the *New Yorker* and invoked the power of the *New York Times,* which had been quite positive about the articles:

> We are awfully glad you are so interested in Rachel Carson's book. . . . You might be interested to know that [of the letters] coming in to the *New Yorker*, 99% of them were for the book and 1% against it. I think the *NY Times* has expressed most people's point of view in their editorial last June. I am enclosing a copy of it and underlining the part that will answer your question about only one side of the story. . . . Mr. Brooks and I both feel the *NY Times* editorial will answer your question. Incidentally, I am rather curious to know who the scientists are that you refer to. Have you a copy of any statement they have made, and if so, would you be willing to let me see it?[99]

The ploy of asking for specific references was used frequently in Houghton Mifflin correspondence with nonsupporters, which had the dual effect of putting them on notice that their claims would be challenged and underscoring the significance of *Silent Spring*'s own supporting documentation.

Ford's personal involvement showed up in memos to Rodell and Brooks, peppered with exclamation marks and references to "our side" and "their side," and in her participation as spokeswoman for the cause on a favorite radio program.[100] Her job, of course, was to keep all media aware of the book, and she kept track of the radio discussions as best she could through her network of informants and clipping services. A very early riser, she happened to catch one such discussion on a predawn farm program broadcast over Boston's station WEEI. Her letter to its host reflects her own vigilance as well as the typical Houghton Mifflin response to media discussion:

> I was very interested to hear your comments on the radio this morning. I am glad you were able to quote some of the favorable comments about Miss Carson's book and also glad you added that McLean [undoubtedly Velsicol's Louis McLean] was an employee of a chemical company. Many scientists, naturally, are employed by chemical companies who scarcely have an objective point of view on all this! I am under the impression the article you read is part of a booklet that is used as propaganda particularly by the chemical companies against the book. Many people that attack Miss Carson's book have not read it and give out such ridiculous statements as she is against all chemicals. She is against their misuse and abuse and anyone who has read the *whole book* knows that. It is far from the truth to accuse her as some of these scientists in the pay of chemical companies are doing.
>
> I thought you would be interested in a piece from last Sunday's *Wash Star*. I know you can't keep on forever with comments on *Silent Spring*.[101]

Houghton Mifflin also used coverage in the media itself as the content of advertising and publicity material. Advertisements sometimes showed an array of clippings spread across a page ad, with the tag "The whole country is talking about *Silent Spring*." Others referred to widespread media comment in "front page coverage from coast to coast."[102] Such ads simultaneously conveyed the ideas that the book's message was a "hot" topic and that it was well received by experts and public alike.

Meanwhile, the chemical industry's campaign against Carson and *Silent Spring* was itself vigorous and multilayered, as I discuss more fully in the next chapter. Someone alerted Brooks that the National Agricultural Chemicals Association had come up with a $250,000 public-relations war chest devoted to *Silent Spring*,[103] an intimidating figure for 1962 and one that likely exceeded Houghton Mifflin's promotional budget by quite

Advertisement for *Silent Spring*. Yale Collection of American Literature, Beinecke Rare Book and Manuscript Library.

a bit. The counter–*Silent Spring* campaign generated two documents in particular that in Houghton Mifflin's estimation demanded response. Monsanto created a parody, "The Desolate Year," for the October issue of its corporate magazine,[104] which its public relations department made available as an advance reprint to five thousand editors, reviewers, and syndicated writers, hurriedly distributed in galley-proof form to meet the book's publication date.[105]

More serious in tone and somewhat more widely circulated, a possibly greater threat was *Fact and Fancy: A Reference Checklist for Evaluating Information about Pesticides*, a booklet cobbled together over the summer by the NACA.[106] Certain passages from the book—directly quoted but neither identified as coming from the book nor credited to Carson—were juxtaposed with "refutations" credited to experts friendly to the NACA. Because the passages were direct quotes without credit or citation, Brooks wanted to sue for plagiarism; but he was advised by legal counsel that such a suit would probably fail. He wrote to the NACA demanding that passages from *Silent Spring* be properly credited to Carson, which the association readily agreed to do in any future printings,[107] an empty promise since so many copies had been printed that reprinting was unlikely. The booklet was circulated widely to editors, reviewers, educators, agricultural interests, government officials, and the like.[108] Brooks told Thompson, "I'm inclined to think it would be a mistake to ignore this publication completely. . . . I think we should call them on this." By "calling" them on the attack, Brooks intended more than demanding proper credit in the booklet; he wanted a public response, prepared by Houghton Mifflin, to ensure that rebuttal against the book's critics would be aired.

Fact and Fancy prompted Houghton Mifflin to undertake its own ten-page promotional booklet mirroring the format of the NACA booklet but much more carefully organized and concertedly documented. Although the project was originally discussed internally in terms of direct response to *Fact and Fancy*, eventually Carson and others expressed concern that it not resemble the NACA booklet too much. *The Story of "Silent Spring"*[109] traced the career of the book, its publicity, and attacks on it. Subheadings indicated the intended direction of the booklet's argument: "What the Book Is About," "Is the Author Qualified?" "The Effects of Pesticides on Wildlife," "Effects on Human Health," "Are We Being Protected?" "Is It All or Nothing?" "How Widespread Is the Danger?" and "What Next?"

The booklet's style was, as the advertising department had planned, "almost as though there were a running conversation going on between the book's detractors on the one hand and its advocates on the other,"[110] but the booklet obviously supported Carson and her message. Approval was given for 100,000 copies of *The Story of "Silent Spring"* to be sent to all on Houghton Mifflin's standing lists of recipients (reviewers, literary magazines, educators, retailers, etc.), as well as to libraries of the "15 most populous states," and heads of wildlife and Audubon clubs. A memo from Brooks directing circulation of the booklet called "newspaper, editorial people important, paramount," adding, "Pitch . . . to future because of forthcoming legislation."[111] Brooks was referring to what he expected, or at least hoped, would be the legislative upshot of congressional attention to the articles and the book.

Houghton Mifflin's booklet was, however, much delayed by in-house deliberateness and Carson's own painstaking review of it, performed slowly as her health continued to fail. Despite hopes that the booklet would be ready before the end of February, revisions and reconsiderations dragged on until it became apparent that there was danger it would not be ready for distribution before the pivotal *CBS Reports* program. A March 13 press release had announced that the broadcast was now firmly scheduled for April 3. Houghton Mifflin had a short time in which to get the attention of television reviewers. Postcards were hurriedly sent out noting the airdate and promising that a complimentary copy of the book was on its way: "We hope you will be able to read it before you view the show."[112] But to Ford's consternation, Carson's supervision and Houghton Mifflin's many revisions continued to delay completion of the *Story* booklet. With little time to spare, she frantically argued for and got permission to distribute it in draft form to television-page editors.[113]

Renewed interest in *Silent Spring* after the television broadcast was gratifying to Brooks and his colleagues, but the administration's PSAC report was yet to come. The positive attention from the CBS program was cause for celebration, but a negative report could prompt a news flurry chilling reception for Carson's message considerably and threatening the long-term viability of what had been one of the most successful books in Houghton Mifflin's history. When the report was released in mid-May, Brooks was elated that its findings fell on the side of vindicating Carson's arguments, and when she was summoned to testify at the two congressional hearings, everyone at Houghton Mifflin involved with the book shared Brooks's excitement.

Presented with the opportunity for the most positive publicity yet, Brooks was inclined to flood the media with a new advertising wave; and ads that reproduced or sometimes mimicked newspaper clippings about the report were placed around the country.[114] He wired more than a hundred of the largest bookstores: "President's report completely vindicating *Silent Spring* calls for immediate action. Hearings scheduled in both House and Senate will keep book before public for weeks to come. Series of reminder advertisements is scheduled in your local paper. Please check stock and display *Silent Spring* where your customers can see it."[115] According to Brooks, "About 10 percent of all the money that comes in on *Silent Spring* seems to be going right out again in space advertising."[116] That heavy investment, however, worked as much to emphasize substantiation of the book's credibility in the face of public challenge as it did to stimulate sales.

Brooks's impulse to capitalize on news events was part of his responsibility as manager of the book's career, and his personal beliefs unquestionably magnified his professional enthusiasm. But Thompson, in a rare expression of his own philosophy on publicity, advocated restraint while the report's content and the congressional hearings were still front-page news. "Publicity such as Rachel is getting breaks all the rules. It's nice sometimes to make a good review and a good ad coincide, but when we are getting front-page space time after time, advertising at the same time is almost an impertinence." He closed the memo with an extraordinarily odd comparison: "We may be wrong. This situation is, after all, unprecedented. The only parallel I can think of is a book as evil as Rachel's is benign. MEIN KAMPF used to have a new wave of sales whenever Hitler invaded a new country. People were trying to see who would suffer next. Here again, the reminder ad seemed most effective!"[117]

In the case of *Silent Spring,* Houghton Mifflin's corporate mission certainly began with the objective of financial and institutional gain through sale of a important book and rights to it—"important" in the sense that it would sustain Houghton Mifflin's cultural and professional status while strengthening the future of the company's prestigious backlist. However, the task of getting the book out and spreading the message was rapidly exceeded by the need to support and defend it in the face of various challenges. There were challenges to publication itself, to promotion, and to the very credibility of author, book, and publisher. But Houghton Mifflin's care to preserve institutional credibility per se was largely eclipsed by a sense of mission that those involved had come to share with their author (and to an extent not seen at the *New Yorker*).

Book versus Magazine

Although the appearance of *Silent Spring* in both book and article form offers a textbook comparison of the two forms, neither Shawn nor those at Houghton Mifflin were likely to make explicit their own views of the differences between the two media. A few comments, however, suggest something of the self-perception of each compared with the other form, as well as the significance of differences between the two.

Shawn made no public pronouncement about the relationship between his magazine and the books that he excerpted, although he reportedly became quite excited about those pieces that had gone on to be books, and he liked to brag that they had.[118] For him, symbiosis between the two

media likely went beyond promotional advantages to a sharing of esteem. But his purpose in publishing the *Silent Spring* series was first and fully to provide *intellectual news* in the form of an exposé of an environmental problem. He had "scooped" the book yet retained the implicit imprimatur conferred by the forthcoming book. As it happened, the journalistic organization of the articles—starting with sensation and delaying to the very end Carson's declaration that she was not prescribing a total ban—was probably responsible at least in part for the intensity and swiftness of the chemical industry's reaction.

Similarly, the Houghton Mifflin staff rarely commented on their relationship with the *New Yorker*, at least not on the record, although indirect references suggest nothing other than cordial if very limited interactions. But publication and promotion plans had to work with and around the fact of prior exposure in the *New Yorker*. The idea that the book, not the articles, was the primary and essential carrier of Carson's message was inherent in much of the Houghton Mifflin discussions, just as it was in Carson's discussions of the project. That priority in status if not chronology was the subtext of an exchange between Ford and Thomas Horgan of the Associated Press about an agreement for an exclusive press release in early June, just before the *New Yorker* articles appeared. Ford proposed the arrangement in a letter gently implying that the *New Yorker*'s involvement had come *after* Houghton Mifflin's decision to publish the book, rather than simultaneously: "In October we publish a shocking book by Rachel Carson entitled *Silent Spring*. *The New Yorker* is so keen on it they are using portions of it in several issues beginning in the June 16th issue."[119]

The fact of prior exposure in the *New Yorker* also limited the timing and extent of Houghton Mifflin's prepublication publicity. In April of 1962, Ford wrote Rodell urging restraint in publicizing the book before the articles had appeared:

> I think it better to let the *New Yorker* come out and let people "discover" this book just as they did with the *Sea Around Us*. I am fearful lest we suddenly become a group with a Cause and get the lobbyists for the paper and chemical companies on our tail with their advance publicity. We do not want to appear as amateur lobbyists. Rather as publishers of a great book by a distinguished author.[120]

Ford's language shows professional loyalty, but it also reveals a view of how book publishers should behave, even those with a "Cause." Although she fully expected the audiences for the *New Yorker* and the book to overlap, it was the magazine's role to uncover a timely topic, while the book publisher's role was to provide "a great book by a distinguished author." Thompson had displayed a similar attitude in seeking Carson's "fiery heart" as the unifying force for the book. That "fiery heart" required care: "We are beginning to edge over into the field of the hysterical protest and that we should leave to the magazines and the editorial writers."[121] Presumably, Thompson was not including the *New Yorker* among that group, but his implied distinction between the objectives of book publishers and that of other print media is noteworthy.

Another issue, more concrete and perhaps more significant, was the difference between a book and a magazine article from the point of view of supporting references. When *Fact and Fancy* prompted Brooks to recommend writing a response to the NACA, he wrote Thompson, "If we can get our letter printed, it would be good advertising with the emphasis where we want it most, i.e., on her documentation, which is in the book, but not in the *New Yorker*."[122] In this context, it is significant to observe one editorial quirk in *Silent Spring* that reflects just how important the weight of that documentation was to Brooks and Houghton Mifflin. The fifty-five pages of notes are, in fact, somewhat inflated by space taken up in the repetition of the word "page" and number (e.g., "Page 158") between entries, effectively adding approximately three to six lines per page of references, which over the entirety of the bibliography could have made a difference of between four and ten pages. This design decision almost certainly was Brooks's; and there may have been a logical or aesthetic argument for the repetition, in that some citations were listed for a span of pages ("Pages 161–62") while others were listed for a single page, which could have resulted in unattractive or even confusing notation. Yet the repetition of, for example, "Page 158" six times between six references seemed suspiciously excessive to an eagle-eyed spokesman for the opposition, William Darby.[123] Had Brooks been asked to shorten the book for some reason, this section would have been the easiest place to carve out a considerable amount of space and text, under normal circumstances. But these were not normal circumstances.

Risk, Symbiosis, and Independence

Obviously, each of Carson's two publishers expected symbiotic benefit from the other's association with *Silent Spring*. For Houghton Mifflin, the *New Yorker* series provided controlled prepublication publicity in an organ

Page 157

Perry, *Gypsy Moth Appraisal Program.*

Page 158

Worrell, "Pests, Pesticides, and People."

Page 158

"USDA Launches Large-Scale Effort to Wipe Out Gypsy Moth," press release, U.S. Dept. of Agric., March 20, 1957.

Page 158

Worrell, "Pests, Pesticides, and People."

Page 158

Robert Cushman Murphy et al. v. *Ezra Taft Benson et al.* U.S. District Court, Eastern District of New York, Oct. 1959, Civ. No. 17610.

Page 158

Murphy et al. v. *Benson et al.* Petition for a Writ of Certiorari to the U.S. Court of Appeals for the Second Circuit, Oct. 1959.

Page 158

Waller, W. K., "Poison on the Land," *Audubon Mag.*, March–April 1958, pp. 68–71.

Page 159

Murphy et al. v. *Benson et al.* U.S. Supreme Court Reports, Memorandum Cases, No. 662, March 28, 1960.

Page 159

Waller, "Poison on the Land."

Page 160

Am. Bee Jour., June 1958, p. 224.

Page 161

Murphy et al. v. *Benson et al.* U.S. Court of Appeals, Second Circuit. Brief for Defendant-Appellee Butler, No. 25,448, March 1959.

Page 161

Brown, William L., Jr., "Mass Insect Control Programs: Four Case Histories," *Psyche*, Vol. 68 (1961), Nos. 2–3, pp. 75–111.

Pages 161–62

Arant, F. S., et al., "Facts about the Imported Fire Ant," *Highlights of Agric. Research*, Vol. 5 (1958), No. 4.

Sample page of "List of Principal Sources," *Silent Spring.*

that could only add to the perceived integrity of the work because of the magazine's special status within the American media system. For the *New Yorker*, Houghton Mifflin's forthcoming book provided weighty intellectual and scientific validation not normally available to the editor of a nonacademic journal—validation implied before publication and actual thereafter. The risk in publishing *Silent Spring* to each firm was reduced by the participation of both, especially as each successfully thwarted efforts to delay or stop publication.

Yet the risk was already reduced by a quality these two publishers shared and contributed to the public life of *Silent Spring*: relative independence. As a publishing house still privately held, Houghton Mifflin had to answer only to its own idea of corporate mission. The *New Yorker*, unique among mainstream periodicals, had the luxury of similarly lofty priorities. This is not to assert, naively, that neither was concerned about profitability, since their ability to remain independent of commercial, political, or social pressures depended on maintaining a comfortable profit margin. But they were operating at a moment in American publishing history when commerce and ideology had yet to become as entangled and agonistic as they would be in later decades with the rise of, among other things, conglomerates. Indeed, by the early sixties a revived form of sociopolitical idealism was becoming rather marketable, and *Silent Spring* is often cited as one of the pioneers of that idealism.

Moreover, the independence shared by Houghton Mifflin and the *New Yorker* played out the author's independence in corporate garb. Both publishers provided Carson with crucial support that protected and insulated her own independence from commercial and political pressure. They also came to share her mission with respect to pesticide misuse, especially Houghton Mifflin. But however unconsciously the publishers may have sensed it, the aura of the author as lone voice was (and is) essential to the role of the book in the American media system; and that aura was essential to what Carson's two publishers shared, took to themselves, and conferred on their respective institutions. They embraced the idea of the individual voice, supported by fact. In doing so, they were observing the very essence of their profession, at least where nonfiction was concerned.

Risk-taking in this context was further part of the lore, indeed the obligation, of publishers who enjoyed such independence. By publishing and supporting the lone author, the *New Yorker* and Houghton Mifflin— Shawn, Thompson, and especially Brooks—were participants and even caretakers of the traditional American ideals of free expression and dissent.

That tradition underlay Thompson's embrace of Carson's "fiery heart" that would stir up a "commotion" now fully public. As for Brooks's gratification at the "super-ruckus," he concluded his memoir declaring that *Silent Spring* "will remain an assurance to writers that, in our overorganized and overmechanized age, individual initiative and courage still count—that change can be brought about not by incitement to war or violent revolution, but rather by altering the direction of our thinking about the world we live in."[124]

CHAPTER 4

Opposition: "How Do You Fight a Best-Seller?"

Carson, her agent, her publishers, and her supporters all knew that her warning about pesticide abuse would not go unanswered. In fact, they counted on the rise of public debate—Thompson's "commotion" or Brooks's "ruckus"—although probably not one with the size, vigor, or acrimony that ensued. Carson had targeted a number of culprits in academic science, government, and industry, so a variety of reactions was predictable, including a formal public-relations campaign to discredit Carson's message. In fact, in the sparse literature on the history of public relations, the *Silent Spring* episode provides a classic demonstration of how to, and how not to, deal with a challenge to an entire industry.

For reasons to be explored, the fact that *Silent Spring* was likely to be a best-seller was of special concern to those Carson criticized. Her objectives—to inform and mobilize—had direct counterparts in what her opponents hoped to do—to counter-inform and fend off mobilization. A number of writers in environmental science and communications have examined their response to *Silent Spring*,[1] and only a sketch of the complex themes and ploys is possible to include here. More important for this story is the context of the public forum in which that rhetoric was applied: the identity and goals of the opposition; the chronology, scope, and form of their efforts; and the significance to them of having Carson's message appear in book form.

Who the Opposition Was and What They Worried About

Essentially, Carson faulted three groups for the hazardous overuse and misuse of pesticides: (1) the academic scientists investigating pesticides; (2) the government agencies that set pesticide policy, supervised their use, and even used them themselves in broadscale projects; and (3) the chemical industry that produced and promoted the pesticides. All three groups rose to the debate but the first two did so less publicly, deferring to their allies in the third. From their perspective, Carson had launched a unilateral attack. Where Carson saw and tried to redress a drastic imbalance in research, policymaking, and public information, her opponents saw a gratuitous and dangerously one-sided assault demanding rebuttal.

Academic scientists, Carson had argued, were culpable for limiting their study to the effectiveness of pesticides, ignoring the broader impact on nature and humanity; and she attributed the narrowness of their focus to the fact that research funding often came from industry. Although her judgment of these researchers was comparatively gentle, according to some observers the academic scientific community nonetheless felt that its professional integrity had been impugned.[2] If peers had contended that a conflict of interest existed in commercial subsidy of their research, their standing in the scholarly community would have been compromised. Conceivably, they might also have feared the loss of that subsidy if their industry partners became shy of apparent connection; but the reality that academics provided valuable and inexpensive external research for industry shielded them from that likelihood. Carson's calls for added research could in fact benefit this group if funding were increased through congressional or executive action.

With a few exceptions to be noted, members of this group generally confined themselves to voicing their concerns in the traditional modes of scholarly communication—the professional and academic journals—or in specialized forums to which they had been invited, sometimes as supposedly objective arbiters between Carson and the chemical companies. The few exceptions were so closely allied with the other two groups (business and government) that they spoke more for those groups than they did for academician-researchers—ironically demonstrating one of Carson's points.

Government agencies, primarily the USDA but also the Department of the Interior, had drawn far sharper criticism in *Silent Spring* than the scientists.[3] Carson's special ire was directed at the "eradication" programs administered by the USDA and at what she felt was the Agricultural Research Service's (ARS) grave skew in favor of copious, unexamined use of pesticides. In fact, both the USDA and the ARS within it had been under scrutiny before *Silent Spring*'s publication; and in 1960 a Committee on Pest Control and Wildlife Relationships was formed under the auspices of the National Academy of Sciences–National Research Council (NAS-NRC), a nonprofit, independent group of scientists drawn from both industry and academe.

Both at the time and much later, observers raised questions about a serious imbalance in favor of industry on this committee. The names of two committee members would appear frequently in the *Silent Spring* debate: I. L. Baldwin, professor of agricultural bacteriology at the University of Wisconsin and chairman of the committee; and George C. Decker, economic entomologist for the Illinois Natural History Survey, chairman of the Subcommittee on the Evaluation of Pesticide-Wildlife Problems, and reportedly consultant to various chemical companies.[4] Baldwin would later review *Silent Spring*, negatively, for the American Association for the Advancement of Science's *Science* magazine;[5] and Decker's "Pros and Cons of Pests, Pest Control and Pesticides," originally appearing in a British journal, was quickly reprinted in the American Chemical Society magazine, *Chemical and Engineering News* (*C&EN*).[6] Both articles were eventually reprinted in quantity and circulated to the media and the public, along with certain other pieces selected by the chemical industry.

The NAS-NRC committee report[7] was published in 1962, most of it appearing prior to the appearance of *Silent Spring*; and it met considerable disapproval from environmentally concerned biologists such as Clarence Cottam (himself a member of the committee) and Roland Clement, both national officers of the Audubon Society, and ecologist Frank Egler (all three were Carson's friends and colleagues). Only the first two of the report's three parts were published because of the refusal of some members, including Cottam, to sign the third part, which they felt omitted the most serious concerns about pesticidal side effects and hazards. Egler wrote in the *Atlantic Naturalist* that the reports "cannot be judged as scientific contributions. They are written in the style of a trained public relations official out of industry, out to placate some segments of the public that were causing trouble."[8] Nonetheless, the two-thirds of the report would later be cited repeatedly as objective "third-party" scientific proof that no problem of the scope outlined by Carson actually existed.

Employed representatives of the Department of Agriculture were not nearly so visible nor so vocal, despite the sharpness of Carson's criticisms. Ernest G. Moore, an employee at the time of the debate and later author of a thoughtful history of the Agricultural Research Service, described the hesitancy and deliberation with which the affected governmental agencies approached the publication of *Silent Spring*, which they clearly felt constituted a crisis.[9] Secretary Orville Freeman was first alerted to the *New Yorker* articles on July 11, 1962, by Byron T. Shaw, administrator for the ARS, who felt that the articles were potentially damaging and "should not be dismissed lightly."[10] A calming note was sounded the next day by Rodney Leonard, a special assistant to the secretary, who felt that (off the record) Carson's work made some sense, if somewhat overstated, and that Shaw's statement was a "typical bureaucratic response which never quite joins the issue and, further, doesn't recognize the emotional impact which the Carson articles have on people." Leonard added, "I fear that if his general attitude sets the tone of the Department's response there will be a strong adverse public reaction." He noted that Shaw had missed the point that Carson was not arguing for complete abandonment of pesticides and asserted, "I think we should recognize that no matter how we might argue against the conclusions in her articles—and in her forthcoming book— we will not convince the public that she is wrong. . . . I believe it would be much better to recognize the need for more research as well as the fact that we are dealing with deadly compounds in using these chemicals. I think the Department has started off in this direction, and her efforts may speed up this effort."[11]

It took a few weeks of internal wrangling through meetings and memos to arrive at an official policy, during which time evidence of the conflicting impulses within the department surfaced in the media. Moore, then public spokesman for the ARS, sharply criticized Carson's articles to a reporter for the *Washington Daily News*, which occasioned internal discussions with Leonard and Shaw to the effect that an official policy statement was needed immediately.[12] Noted the *Wall Street Journal*, "Freeman squelches trigger-happy underlings who itch for quick rebuttal of Rachel Carson's magazine attacks on safety of chemical insecticides. The Agriculture Department builds a careful defense of its encouragement of insecticide use. An indirect reply: the Department pushes work on non-chemical war against bugs."[13]

Eventually the USDA's official stance—imposed in part by the Kennedy administration in a July 31 meeting with White House Science Adviser Jerome Wiesner—was, indeed, to emphasize what research had already been done and the positive value of continuing if not increasing research.

But the unofficial strategy was to say as little as possible on the record beyond that, even at the time of the release of the President's Science Advisory Committee report in May 1963. That constraint rankled those within the USDA considerably.[14] Nonetheless, Shaw provided the conservative *U.S. News & World Report* with an extended interview during which he reassured the public and defended the use of pesticides.[15]

Meanwhile, the USDA was in frequent contact with representatives of industry, undoubtedly with the idea of closing ranks. Public-relations efforts on the part of private industry could serve some of the USDA's purposes, allowing government spokespersons to take a much lower profile, forced as they were to deal with response from several quarters within the government and from the public at large.

What was at stake for the USDA was first the esteem and confidence of the public, imposed indirectly but forcefully by congressional and White House "pass-through" pressure. Of the hundreds of letters from citizens about *Silent Spring* archived in USDA records, a substantial number were forwarded from the White House or from a Capitol Hill office with cover letters expressing concern and necessitating replies from the department, not only to the constituent letter-writer but also to the senator, representative, or White House officer who had first received the letter. Thus, by extension, the jobs of those in the ARS or the USDA in general depended in part on their handling of the situation with respect to *other* arms of government. In the long run, however, pressure to increase research offered the possibility of actually improving their lot through an infusion of funds for new tasks, as Leonard and others most likely understood. The situation of departmental employees who already worried in silence about the hazards of pesticides or chafed under industry dictates might, furthermore, become more comfortable.

Yet there were also those at the higher levels of the department who came from industry and had strong industry ties; others perhaps anticipated retiring from civil service into industry. What was at stake for them was the need to preserve cordial relations with agribusiness and the chemical industry. The fine line between official restraint and accommodation of the private sector could be crisscrossed behind the scenes; and cooperation between industry and government was best directed at controlling the public discussion insofar as possible.

The response of other governmental departments, such as the Food and Drug Administration (FDA), Public Health Service (PHS), and Fish and Wildlife Service, was much less evident, plausibly because they had a sense less of crisis and more of opportunity. Moreover, what was at stake for them related, first, to their relationship to Agriculture. Calls to reorganize agency responsibility for pesticide policy, testing, and regulation were already being heard, in hopes of easing Agriculture's domination of the entire matter. Part of the criticism of the NAS-NRC report concerned its failure to provide any basis for a much-needed coordination across several agencies, which would have reduced Agriculture's ability to act unilaterally without restriction in eradication and similar programs. The July 31 White House meeting under Wiesner had included heads of all involved agencies, and Agriculture's displeasure at instructions not to make any direct counterattack on *Silent Spring* may well have been related to potential loss of clout compared with that of the other agencies. That discomfort was only exacerbated when Kennedy named an independent PSAC task force on pesticide abuse, not to mention when Agriculture officials were summoned to congressional hearings following release of the PSAC report. Thus the FDA, the PHS, and the Department of the Interior stood most to gain by standing by and saying as little as possible beyond expressing reassurance and interest in future research, even if that meant greater controls on pesticide use.

Finally, it must be acknowledged that any arm of government was obligated to worry about the possibility of public panic or hysteria, which at the very least could upset established order and patterns and might conceivably involve threats to public health—if, for example, eating habits were so drastically changed that good nutrition was endangered, or if disease-bearing insects were somehow allowed to overrun the country unchecked. Still, even in light of recent encephalitis outbreaks, experienced higher-level administrators in the USDA or other departments were unlikely to view the prospect of such large-scale problems imminent or even probable, yet they often debated the nature of the crisis they faced in just those terms.

In the main, it was the chemical industry—unfettered by the professional or political constraints imposed on academics and government—that provided public opposition to *Silent Spring*, followed closely by those in agribusiness (sometimes both being represented by the same people). Their counterattack was the most organized, vigorous, and vociferous of all. And obviously, they had both the motivation and the financial wherewithal to mount an extensive campaign. According to E. Bruce Harrison, hired as "Manager of Environmental Information" (along with the New York public-relations firm of Glick and Lorwin) by the Manufacturing

Chemists Association (MCA) for the purpose of dealing with the *Silent Spring* crisis, the firms most immediately involved were Monsanto, Du-Pont, Dow, Shell Chemical, Goodrich-Gulf, and W. R. Grace,[16] although Velsicol, a smaller chemical producer, also achieved some public visibility through its secretary, Louis McLean, after his communications with the *New Yorker* and Houghton Mifflin.

In addition, American Cyanamid involved itself through the highly visible activities of scientist Robert White-Stevens, who functioned more than anyone else as spokesman for the industry, most memorably on the *CBS Reports* program. White-Stevens went on a national lecture circuit for much of the time between the summer 1962 publication of the *New Yorker* articles and the June 1963 congressional hearings. He was an imposing, authoritative figure who spoke with a British accent and bombastic style. Another Cyanamid connection was Thomas Jukes, who wrote frequently and in flamboyant style to—among many—the USDA, Houghton Mif-flin, and Carson herself, though not identifying his association with Cy-anamid in his correspondence. He may also have been the author of some of the more inflammatory but anonymous pieces written in response to *Silent Spring*: "White-Stevens did most of the talking, I guess, and I did most of the writing," Jukes said in later years.[17] Based on the rhetoric of his own *Silent Spring* parody in the August 1962 issue of the National Agricultural Chemicals Association's *NAC News and Pesticide Review*,[18] one can easily suspect him of being a collaborator if not primary author for Monsanto's widely circulated parody, "The Desolate Year."[19] Mon-santo, in fact, saw itself as spearheading the countering effort. "Two prin-cipal options were considered: to ignore the whole thing and let some other company or a trade association speak up, or to fight back," said former executive Dan Forrestal.[20]

Otherwise, the individual companies rapidly agreed to speak as one voice, allowing the trade organizations, notably the MCA and above all the NACA, to be that voice, often in the person of George Decker. Wrote the *C&EN,* "Most producers of pesticides aren't commenting for publica-tion about the Carson articles, preferring instead to let their trade associa-tions carry the ball."[21] Allied Chemical's response to requests for comment was typical: "We have not issued a statement on use of pesticides but have adopted the general policy as set forth by the Manufacturing Chemists Association."[22] Reports about the $250,000 "war chest" that so worried Brooks soon popped up in industry publications.

Agribusiness, including parts of the chemical industry, was represented by the Nutrition Foundation, a nonprofit but well-supported group whose membership comprised fifty-four companies, both food producers and chemical producers, that were involved in all aspects of food production, including production and use of pesticides. The foundation funded academic research and provided a coordinated public forum for its membership. The president of the Nutrition Foundation, C. G. King, spoke occasionally on *Silent Spring,* but he was more usefully a letterhead name for some of the largest direct-mail efforts. His colleague and spokesman, however, Frederick J. Stare, head of the nutrition program at the Harvard School of Public Health, rivaled White-Stevens in visibility and ubiquity. Stare had already acquired a contentious reputation locally, having been sued two years earlier by the Boston Nutrition Foundation, which alleged slander and libel; and his public contempt for organic "food faddists"[23] added to the energy with which he took on Carson and her book. Yet the relationship between Stare, the industry-funded Nutrition Foundation, and the corporations that supported the group remained obscured in the public discussions, as did collaboration between the Nutrition Foundation and the MCA.

What was at risk for these industries was first—and probably least—sales revenue from pesticides. In fact, Dennis Hayley, information director of the NACA, told the primary advertising and marketing trade journal, *Printers' Ink,* that the book posed no threat to the sales position of the industry.[24] Sixty percent of retail sales were to farmers, who realistically had no good alternative to using commercial agricultural pesticides. Even if any great number of home owners or gardeners (representing another 20 percent of retail sales[25]) had chosen to avoid purchasing pesticides, the agricultural market would have remained. Pressure from farmers to provide safer chemicals was easily met by offering more education as to safe use while further research was conducted. But nearly a year after *Silent Spring* first appeared, industry spokesmen told *Business Week* that retail sales remained strong and were likely to be unaffected by the release of the PSAC report.[26]

More of a concern, however, might well have been the potential loss of revenue from government programs such as those to control mosquitoes and Dutch elm disease, as well as the fire ant and gypsy moth "eradication" programs that Carson particularly excoriated. Although one dollar-amount floated through the debate in an oft-quoted epithet—"the $300-million pesticide business"[27]—the actual value of total production of insecticides, herbicides, and fungicides, including the proportion of government contract revenue, is difficult to determine. Most estimates of agricultural chemical use (including fertilizers) hovered between $700 million and $1 billion per year in that era.[28] If the frequently cited figure of $300 million referred primarily to commercial sales of pesticides, the amount of government purchases beyond that was never stated. Given that these programs were unpredictable and subject to bid, however, the questions of manageability and profit margin were likely problematic for many producers. In fact, some producers publicly blamed government programs for the uproar over pesticide misuse because of the poor management and control of the public spraying programs.[29]

Loss of confidence in food products was also a concern, but again the absence of realistic alternatives limited consumers' options. The possibility did exist that food producers might be faced with a new competitive war revolving around claims about purity and safety. But the logistics and feasibility of changing established food-production and food-processing practices were then daunting and unlikely to be undertaken by the industry on its own initiative, particularly if it believed that the quality of its food was not endangered. Thus public reassurance about food quality was a logical industry response to *Silent Spring*'s warnings about possibly dangerous medical consequences of pesticide contamination.

Worries about prestige overall were reflected in the corporations' public posturing as injured parties or as keepers of a public welfare that they believed would be much endangered if pesticides were banned. *Printers' Ink* itself described the public relations effort as one aimed at recovering the "prestige lost" because of the book.[30] The exact nature of that prestige in the eyes of the general public would have been difficult for anyone to measure, define, or describe; but it was an era in which questions about the worth and justification of the profit motive were being asked in the general press. A series of ads and publications from DuPont, among others, defended the profit motive and suggested that business was a "good career for a college man."[31] Thus the early signs of 1960s antiestablishment sentiment in a young population had already caught the attention of corporate public affairs officers, faced now with defending the good name of their firms against something other than charges of fiscal irresponsibility or occasional criminal malfeasance.

Similarly, a shift in the climate for the conduct of American business was part of the change from the Eisenhower to the Kennedy administration just two years before *Silent Spring*. Although the Eisenhower administration had seen an infusion of private-sector management in cabinet agencies, the Kennedy administration seemed bent on changing some of the rules; and signs of increasing social activism were likely unsettling to business. Despite some continuity in personnel in agencies such as Agriculture, these changes meant at the very least uncertainty for the private sector, particularly as the public seemed increasingly disposed to demand input into policy beyond the ballot box.

It was also, however, an era in which public relations was becoming increasingly professionalized and institutionalized, occupying an ever-more important role in corporate structure. The chemical and agribusiness industries were already availing themselves of public relations expertise to counter mounting questions about the role of science in American culture. Efforts to generate good will were evident well before *Silent Spring* in advertising campaigns like DuPont's "Better Living through Chemistry." Moreover, the need to use other tactics in concert with advertising campaigns was already well acknowledged, and "prestige" was a quality that required both selling and defending.[32]

Industry's greatest worry, however, was the one least mentioned: the prospect of increased regulation, oversight, and control from outside the industry and its governmental friends, something already causing increased concern within trade groups.[33] In a *C&EN* article outlining the industry's purportedly late response to the *New Yorker* articles, a Stauffer Chemical officer made explicit his concern that "if the public becomes frightened about its food supply and general health through 'contamination' by chemicals, governments at all levels will push for unnecessarily increased regulation."[34] This concern was obscured in the public debate by virtually all other arguments, yet it almost certainly represented the essential fear justifying the sizable public-relations expenditures devoted to dealing with *Silent Spring*. New regulations could be far more costly than a decline in nonagricultural purchase of pesticides, and more costly even than a public-relations effort to inform and reassure the public about food and pesticide safety. Executives interviewed by *Business Week* just before the 1963 congressional hearings indicated that they were "particularly concerned over the possibility of legislation so restrictive on new products as to take the profit out of research efforts."[35]

How to Respond: The Options

The first question was obviously whether to respond at all. For some, the initial impulse was aggressive: "First, why can't the industry take off the gloves once in a while in a controversy like this? Have our modern public relationship techniques made mutes or sissies out of all our business executives? Must we be so gentlemanly when someone hits us below the belt? And second, why can't we turn our defense into a counter attack?"[36] The assumption was implicitly that the public would take *Silent Spring* seriously enough to prompt undesirable change—change directly and drastically affecting industry. The greatest risk, therefore, lay in *not* responding. The MCA public-relations director said at the time, "If we don't take charge now, we're going to be buried, pure and simple."[37] And the industry had the proper resources to take charge. "This was, for us, an opportunity to wield our public relations power," said a Monsanto official.[38] For the most part, risks in failing to undertake internal reform within industry went ignored, let alone discussed publicly, though government agencies for their part had no choice but to deal with congressional and presidential calls for greater control and supervision of pesticides.

The unanimity of industry on the point should not be overstated, however. Some voices in the chemical industry suggested that "to engage in a public debate with Miss Carson may only call even more attention to her viewpoints than they might otherwise receive."[39] Indeed, the official approach for both business and government interests, at least initially, was to take a positive offensive touting the necessity and benefits of pesticide use, emphasizing their safety (when "properly used"), and proclaiming research on them as thorough and ever-improving, especially should more funding be made available. As *Printers' Ink* reported, "It has generally been decided to ignore Miss Carson herself and accentuate the positive side of agricultural chemicals. . . . The [MCA, NACA,] and producers of the pesticides wisely decided to let the popular authoress alone and step up educational and informational programs instead."[40]

Similarly, the editor of *Chemical Engineering* wrote that Carson should be commended, not castigated: "Rather than protesting too much, might not we take the questions Miss Carson raises as a helpful reminder of the questions we must constantly be asking ourselves? Is it best for the industry—and for the public—that pesticides be looked upon, anywhere, as friendly household commodities rather than as poisons?"[41] Generally, however, the work of the chemists within corporate walls seemed little affected by the media controversy of the moment, although at least one Monsanto executive later indicated that Carson's warnings had some resonance within the industry: "The new era forced all companies to take a harder look at the way they and their products were impacting the quality of life."

But the same executive said, "I've been asked in recent years whether Monsanto didn't overreact. At the time, it didn't seem so. Actually, I'm glad we spoke out."[42]

The decision to respond had, notably, been catalyzed by the industry's horror at the first, rather dramatic installment of the *New Yorker* series; and the temptation was to come out with guns blazing. Responses varied across a fairly wide spectrum, however, from emphasis on the positives of pesticides to full-scale attack on Carson and her work. Eight publications in particular formed, in effect, the core of the opposition's arsenal. In various combinations, these few pieces were broadly circulated to press and public over the course of the debate. Taken together they indicate the character of the rebuttal.

First was the booklet quickly prepared by the NACA, *Fact and Fancy*,[43] in which unattributed quotations from *Silent Spring* were juxtaposed as "Allegations" with "Facts" credited to experts friendly to the NACA. Shortly thereafter came Monsanto's "The Desolate Year," a vivid pastiche of Carson's opening chapter, using florid language and often revolting images to dramatize the horrors of a world without pesticides. The parody included some graphic descriptions of the world as "the garrote of Nature rampant began to tighten."

> Genus by genus, species by species, subspecies by innumerable subspecies, the insects emerged. Creeping and flying and crawling into the open, beginning in the southern tier of states and progressing northward. They were chewers, and piercer-suckers, spongers, siphoners, and chewer-lappers, and all their vast progeny were chewers—rasping, sawing, biting maggots and worms and caterpillars. . . .
>
> A cattleman in the Southwest rubbed the back of a big red steer, and his hand found two large lumps under the hide. . . . gritting his teeth, he placed his thumbs at the sides of one of the lumps and pressed. The hair parted, a small hole opened and stretched. A fat, brown inch-long maggot slowly eased through the hole. . . .
>
> But food and fur animals weren't the only ones that died to the hum of the insects that year. Man, too, sickened, and he died. . . . One day, he was stricken by an old foe that had returned violently—malaria. While he suffered, the mosquitoes kept biting, and as each keen proboscis siphoned off his blood it also sucked in deadly gametocytes that were in the red corpuscles. . . . [he] suffered the fiendish torture of chills and fever and the hellish pain of the world's greatest scourge.[44]

The third publication came from Louis McLean, Velsicol's secretary and lawyer who had challenged both the *New Yorker* and Houghton Mifflin.

Titled *The Necessity, Value, and Safety of Pesticides,* it was a pedantic, quer-
ulously legalistic, eighteen-page tract, more philosophical than scientific or
practical.[45]

Next, three book reviews were written by insider-defenders of contemporary practices regarding pesticides, also reprinted for distribution. The
October 1 issue of *C&EN* carried a decidedly negative review by William
Darby, a Vanderbilt nutritionist who wrote often for the trade journal (and
an exception to the rule that academics largely confined response to the
scholarly arena). The piece was originally titled "Silence, Miss Carson,"
but as a reprint it became "A Scientist Looks at *Silent Spring.*"[46] Darby
castigated Carson for ignorance, bias, and irresponsibility. Frederick Stare's
acerbic "Some Comments on *Silent Spring*" was identified as coming from
the January 1963 *Nutrition Reviews,* although it was distributed in at least
one information packet early in January. Targeting the quality of Carson's
research, Stare wrote, "I have seen no evidence . . . which justifies calling
Miss Carson a scientist" and emphasized his own academic credentials and
contacts.[47] Finally, I. L. Baldwin's less negative but still critical review from
Science magazine was dated one day after *Silent Spring's* official publication
date, and as a reprint it retained its original title, "Chemicals and Pests."
Although Baldwin, chairman of the NAS-NRC committee on pesticides,
conceded that some abuse of pesticides might occur in practice, he saw
Silent Spring as a unilateral assault and faulted Carson for not listing the
benefits of pesticides or the cost to humanity if they were banned, observing, "Mankind has been engaged in the process of upsetting the balance of
nature since the dawn of civilization."[48]

Because they were probably prepared *before* the *New Yorker* series ap-
peared, two other pieces often enclosed in the mailings referred neither to
Carson nor to her book; and probably for the same reason, they were
rather more on-point on the issues of pesticide use and safety. Dated
October 1962, *Facts on the Use of Pesticides* (the so-called Cornell Report)
was prepared at Cornell University by the New York State College of
Agriculture at Ithaca and the state Agricultural Experiment Station in
Geneva.[49] It offered specific information in question-and-answer format
about the necessity and dangers of pesticide use, concluding with the
recommendation that the public learn more about the subject and support
more funding for research. The second was George Decker's "Pros and
Cons of Pests, Pest Control and Pesticides," reprinted from the initial,
spring 1962 issue of the *World Review of Pest Control*[50] and designed to
answer criticisms already posed prior to publication of *Silent Spring* by

critics such as the National Audubon Society. These latter two publications were sometimes enclosed with official policy statements from government agencies and represented perhaps the mildest part of the spectrum of negative responses to *Silent Spring*.

In fact, between early summer of 1962 and the following spring, a fair amount of material on pesticide use appeared in the popular press that made no reference to *Silent Spring* or its contentions, nor was the material identified as coming from opposition sources. Typically, this material took the form of articles in major magazines or newspapers, with titles such as "The Safe Use of Pesticides" or "The Truth about Pesticide Use." Although it is perilous to assert that such articles originated from outside the given publication, the presence of similar pieces in direct-mail packets suggests that the public-relations effort included preparation, placement, and distribution of generic, apparently non-argumentative material—as well it should have, from the point of view of standard public-relations practice. Similarly, some effective interviews with the less strident representatives of government and industry were confined, by the interviewee, to discussion of benefits and safety in pesticide use, without reference to Carson or her book. The intended effect of this more moderate approach was to create the impression both of greater objectivity and more encompassing concern for public welfare than Carson's.

Many were also quick to defend existing levels of research and regulation. Frequently, publications cited the Insecticide, Fungicide and Rodenticide Act of 1947 and the 1954 Miller Pesticide Residue Amendment to the Food, Drug and Cosmetic Act, and referred readers to several NAS-NRC publications, as evidence that government study and regulations were already in place and reassuringly adequate. One defense of research and testing of pesticides drew on the concurrent media discussion of birth defects from thalidomide. One editorial stated, "If thalidomide . . . had been an insecticide instead of a drug for human use, it would never have reached the American public";[51] and that parallel was drawn in a number of public debates and news stories. Others pointed out that even if more ecologically sound alternatives were in the offing, they were as yet unavailable, too costly, and only sporadically effective.

The most persuasive opposition to *Silent Spring* sought to take the high, positive ground. Their arguments were that there was good reason for chemical pesticide use; that it was possible to use the chemicals conscientiously, safely, and effectively; that controls were adequate but that Carson's call for more public education and research on side effects merited notice.

Baldwin concluded his largely negative evaluation of the book with the note that *Silent Spring* "will serve a useful purpose, if research on better methods of pest control is stimulated and if all concerned with the production, control, and use of pesticides are stimulated to exercise greater care in the protection of the public welfare."[52] Less graciously, one industry vice president declared, "The industry deserves a black eye for not educating pesticide users on the proper use of powerful chemicals. One of the big problems we've always faced is overestimating the intelligence of users."[53]

Many participants in the debate took Carson on much more directly, and among those, some were more respectful than others. Such direct confrontation required one or both of two approaches: either redefining her message as one that should have been a balanced, "objective" piece of scientific research but one that had failed to present the positives of pesticide use; or, more aggressively, depicting Carson's work as fatally flawed science and recasting her position as irrationally and irresponsibly demanding a total ban on pesticides.

The first, more moderate approach—calling *Silent Spring* "one-sided"—was not only more convincing but also provided a platform on which the opposition could command public attention to present its own views. The NACA's official statement used three bullet points in commenting on the *New Yorker* articles:

- They present one side of a very important subject.
- They present neither an accurate nor a complete idea of the importance of pesticide chemicals to the production of food and fiber or to the protection of public health.
- They allow readers to draw conclusions which are unwarranted in the light of scientific facts.[54]

The expectation that a book (or article) about science should be "scientific"—that is to say, disinterested, "objective," and "balanced" between two inferred views—was a rather disingenuous note that Carson's opposition struck often. As Stare said in an appraisal that would be picked up in several newspapers, "Miss Carson writes with passion and with beauty, but with very little scientific detachment. Dispassionate scientific evidence and passionate propaganda are two buckets of water that simply can't be carried on one person's shoulders."[55] (That Stare's own efforts—not to mention those of others among the opposition, especially Monsanto's "Desolate Year"—constituted their own "passionate propaganda" was an irony probably not lost on many newspaper editors.) Calling the book "an impassioned tract," one reviewer did concede that "it was not meant to be a scientific document" but compared Carson to a "pamphleteer" who ignored the rules of proper scientific engagement.[56] This sort of approach thus defined *Silent Spring* as if intended solely as a work of scientific scholarship in which the presence of an advocate's voice discredited it altogether, rather than the "defended truth" offered by a concerned expert-citizen.

In the same vein, much public criticism attempted to discredit *Silent Spring*'s science. For example, Darby wrote, "Miss Carson has effectively used several literary devices to present her thesis and make it appear to be a widely held scientific one. She 'name-drops' by quoting or referring to renowned scientists out of context."[57] As Carson herself suggested to the Women's National Press Club, characterizing scientific reference as "name-dropping" was a way of subverting the scientific validity of her research and discussion. Cover letters sent with the various opposition reprints carried comments like those of Nutrition Foundation president C. G. King: "I am writing you because many responsible citizens, including both scientists and non-scientists, are increasingly concerned about the promotion of Rachel Carson's book *Silent Spring* as if it were a reliable and 'scientific' document."[58]

Once the book appeared and sources could be checked, there were those who criticized Carson's information as out-of-date and not cognizant of modifications made over time. One Dow official asserted that "Miss Carson has indulged in hindsight. . . . In many cases we have to learn from experience."[59] But this potentially legitimate tack was not often taken. More frequently the position was that although her facts may have been correct, she had reached "unwarranted conclusions" from them, as Parke Brinkley, chief executive officer of the NACA, told the *New York Times Book Review*.[60] And Stare skillfully turned criticism of her conclusions to his own use: "Though I disagree strongly with Miss Carson's central conclusions, undoubtedly she has rendered a service by obliquely arousing an apathetic, unscientific public to a most serious problem—providing enough food for an exploding population, and protecting man from the scourge of epidemic disease." Stare's parting shot reappeared in numerous press reports on the growing debate: "Miss Carson could have written a book which would have helped bridge the gap between science and the public, instead of one which widens the gulf."[61]

Robert White-Stevens's criticisms typified the most pungent, and they continued long after the book was published with its documentation and disclaimer fully available. White-Stevens had been quoted as calling the book a "nostalgic picture of Elysian life in an imaginary American village of former years where a harmonious balance of nature prevailed until pesticides were introduced."[62] The phrase "balance of nature" was intended to refer to an irrational state of mind antithetical to good, serious science. In the 1963 *CBS Reports* interview, White-Stevens called the claims of the book "gross distortions of the actual facts, completely unsupported by scientific experimental evidence and general practical experience in the field."[63] This sort of response characterized the book as "fancy," "fantasy, a kind of 'science fiction,' "[64] and it declared Carson's treatment of the topic of pesticides unscientific, distorted, selective, overstated, alarmist, hysterical, and reliant on scare tactics.

Beyond questions about the quality of Carson's science, claims of hysteria and scare tactics resounded through much of the debate, implying irresponsibility while referring to the detractors' own sources as "responsible authorities."[65] The stridency of the attacks often exceeded the level of drama in Carson's own writing. *PR News* itself joined those who faulted Carson's writing as "hysterical, dangerous extremism," while it described Monsanto's exaggerated imitation of her style in "Desolate Year" as admirable, persuasive prose.[66] One or two other critics also inexplicably applauded the parody as calm and measured compared with *Silent Spring*.[67] Yet even though "Desolate Year" was undoubtedly satisfying for Carson's opponents, its intent to undermine the skill of Carson's writing may still have backfired. Even *Printers' Ink* characterized the parody as melodramatic and "luridly written."[68]

In deliberately misrepresenting *Silent Spring* as a call for wholesale abandonment of pest control and threatening resultant horrors of famine and disease, the parody readily illustrated the industry's own "scare tactics." The threat of such dire consequences, were the public to take Carson and her book seriously, was a recurrent theme. Carson's critics warned of more subtle threats as well, including one perverse twist: a warning to the public that if consumers purchased fewer pesticides because of *Silent Spring*, corporations would not have enough revenue to support research into safer products.

Criticism of the author's treatment of her subject was concomitantly criticism of the author; and despite early resolve to avoid taking on Carson herself, some could not resist targeting the person of the author, subtly or not so subtly. The earliest *New York Times* coverage quoted a chemical company president as saying Carson wrote "not as a scientist but rather as a fanatic defender of the cult of the balance of nature."[69] Calling the book "unscientific" was often combined with asserting that Carson was not a scientist, overlooking or denigrating her credentials and emphasizing the credentials of the opposition's representatives in contrast. The "Miss" honorific, which in the era was a required matter of respect, nonetheless allowed them to underscore the absence of "Dr." before her name each time they referred to her in the company of others who did hold the title. White-Stevens used this device to particular advantage by adding a slight sibilant hiss to each utterance of the phrase "Miss Carson" in the *CBS Reports* interviews. Similarly, Stare's insistence on noting all four of his degrees set up an imposing contrast with Carson's two, which were often not even acknowledged. [70]

Using Carson's gender against her was a possible ploy in the America of the early sixties, but doing so required a degree of care. Former Agriculture Secretary Ezra Taft Benson's question as to why "a spinster with no children is worried about genetics"[71] was quoted to many, but it did not appear in the major media. In his editorial for the *Archives of Internal Medicine*, however, William Bean wrote, "*Silent Spring*, which I read word for word with some trauma, kept reminding me of trying to win an argument with a woman. It can not be done."[72] Others have considered gender issues with respect to the *Silent Spring* debate at some length, including the sexist aspects of criticism of Carson,[73] but misogynistic attacks were rarely as overt as Bean's. Most often a term like "hysterical" was a euphemism for the idea that, as a woman, she was not capable of objective reason or competent science. Conceivably for the very purpose of fending off the appearance of misogyny, one authority to whom Carson's critics turned was (Dr.) Cynthia Westcott. Westcott was the one woman on the NAS-NRC committee, and she had "vigorously defended that committee's report." Her negative review of *Silent Spring* was included in some mailings.[74] Perhaps for health reasons or perhaps to avoid the appearance of a catfight, Carson declined an opportunity to debate Westcott in person, which invited speculation that she could not have held her own against a woman who was a more credentialed or "real" scientist.

One particular point of contention between Carson and Westcott, however, was the issue of financial motivations in pesticide research;[75] and the issue of Carson's own motives constituted another aspect of her opposition's attempts to discredit her. Carson's criticism of private funding of academic pesticide research had struck a nerve particularly within the academic community, and her suggestions about potentially venal corporate-government alliances provoked members of industry (and likely those in government as well). Perhaps as a way of responding "in kind," Nutrition Foundation president King's mass mailing cover letter asserted that the enclosures (including Darby's review) were "written by men of . . . stature as scientists who are familiar with the subject matter of the book, and likewise are in independent positions and widely recognized for their integrity as public-spirited citizens." In contrast, he deplored the fact that "the sales of *Silent Spring* are benefiting the many special interest groups who are now feeding on its publicity. These groups, who are profiting from the atmosphere it attempts to create, are those who advertise and promote to gull and frighten Americans. . . . Their alarmist advertising is aimed right at our pocketbooks. . . . They poison minds for profit, and Miss Carson's book has become a tool in their hands."[76]

Outright claims that Carson's goal was solely monetary had limited credibility, though the suggestion did surface in various forms. Anne Ford's protest to a Boston farm radio program focused on the point, complaining that the show's hosts had made "some rather uncharitable remarks, it seemed to me, about Rachel Carson's writing *Silent Spring* just for the money. . . . Actually your inference was very much the same as the line that some of the chemical company propagandists have been taking. Maybe you didn't know this."[77] Some correspondents directed that charge more credibly against Houghton Mifflin, which was, after all, a profit-making business. Others—including the prolific letter-writer Thomas Jukes—occasionally asked whether it was more acceptable to make money on book sales than on chemicals.

Inevitably, there were those who suggested that Carson had questionable political motives; foremost among them was Velsicol's McLean. His letters to Houghton Mifflin and the *New Yorker* included Cold War references to "sinister influences, whose attacks on the chemical industry have a dual purpose: (1) to create the false impression that all business is grasping and immoral, and (2) to reduce the use of agricultural chemicals in this country and in the countries of western Europe, so that our supply of food will be reduced to east-curtain parity."[78] His article as rewritten for a mailing enclosure stepped back from so clear an accusation of Soviet-bloc sympathy; but in questioning the hidden forces prompting California governor Pat Brown to appoint a committee to investigate pesticides, McLean wrote, "There is no intent to suggest that any person named or referred to by indirection in this writing is a fellow-traveler. The fact is, however, those sinister influences do exist. Where do some small naturalist organizations and individuals find funds to cover nation-wide distribution costs?"[79] In a patriotic attack on those "who belittle the United States farmer and the manufacturers of agricultural chemicals," he challenged the critics: "If they are not ignorant, what are their motives?"[80] Although a few other letter writers raised similar questions to the *New Yorker*, Houghton Mifflin, and elsewhere, this issue rarely surfaced in the media-borne debate beyond the pinkish tint in the word "propaganda."

Rebuttals to *Silent Spring* thus ran the gamut from defense of pesticides to attacks on Carson's motives. But regardless of response, her critics had the common view that a first shot had been fired by Carson; most believed that a response to *Silent Spring* was warranted and that it needed to be substantial. The scale of their reaction might have seemed curiously excessive had Carson's message come in a mimeographed broadside handed out by a garden club member or even a small booklet written for the Audubon Society. The opposition's reaction seemed to be fueled by the doubling of the perceived attack on them, first from the *New Yorker* series and then from the Houghton Mifflin book. But why should publication in a limited-circulation magazine and in a book that might be purchased by only a few hundred thousand people be so worrisome?

Book versus Magazine

Once again, one must remember that in the public forum, the *New Yorker* articles were the first exposure of Carson's message. The relatively vivid initial installment jolted her opponents—the chemical industry above all—into action. Monsanto's Dan Forrestal noted, "*New Yorker* Magazine was the trigger. Without an assist from the *New Yorker*, Monsanto might very well have adopted a safe and cautious course of not making waves."[81] The opposition's first responses were conditioned by two aspects of the first installment: the vivid parable and the absence of Carson's disclaimer that she did not advocate a complete ban on pesticides. More important, they knew that the magazine was read by influential people, including others in the media, and they knew that the news media might take note of something deemed important enough to serialize. Above all, they knew that what they saw in the *New Yorker* would soon be presented in a well-publicized book.

The fact that Carson's treatise would appear in book form raised special warning flags. Parke Brinkley, head of the National Agricultural Chemicals

Association, wrote to Agriculture's Rodney Leonard after a meeting: "I think Miss Carson has set up a potentially dangerous chain of events. While I am not particularly concerned about her book itself, I am concerned about the many other pieces that will be written as the result of it."[82] The prevailing view was that a vigorous and timely response was advisable, in part because the book could be expected to be a best-seller "and every major magazine and newspaper will feel compelled to review it."[83] "It is going to galvanize all the fanatics . . . and the chances are the tumult will attract the attention of editors and cause considerable public comment," wrote DuPont's press analyst to top management.[84]

Thus, dealing with *Silent Spring* was not going to be a mere matter of issuing press releases on Carson's arguments in the *New Yorker* and simply pronouncing her wrong. Even if the *New Yorker* had a policy of publishing letters or rebuttal articles (which it adamantly refused to do), Carson's critics had a far larger challenge: "How Do You Fight a Best-Seller?" asked *Printers' Ink*.[85] Doing battle with a book, especially a potentially best-selling one, entailed matching and even mimicking its influence via the "many other pieces that will be written," beginning with reviews and editorials. From the start, Carson's opposition sought to match if not outdistance the reach provided by the book form and compounded somewhat by appearance in a national magazine.

However, doing battle with a book also meant confronting the inherent weightiness of the book form, including both the intrinsic authority of Carson's research and the extrinsic authority of her publisher's support. With few exceptions, Carson's critics were not conscious of or explicit about the special problems created for them because *Silent Spring was* a book. But one can divine from their efforts a number of attempts to parallel that authority, without—notably—publishing a book themselves.

First was the matter of scientific documentation. Even before the book appeared, the expectation was that it would include supporting documentation not included in the *New Yorker*. Several accounts reported that the NACA and member companies scrambled to put staff to work arduously dissecting each instance of pesticide harm or abuse described in the articles, without benefit of citations to pursue and check. "Ten or 12 companies are preparing point-by-point rebuttals to the *New Yorker* series for their own use. These will take time, mean many man-hours of literature searching and writing," reported *C&EN*.[86] DuPont's July request for review copies, which had so unnerved Rodell,[87] was undoubtedly as much to obtain bibliographic material as it was to compare the scope of the book's charges with the magazine articles. The point was to get an early start on challenging Carson's science, something likely to be more impressive in the book than in the articles.

To frame *Silent Spring* as "failed science" successfully, much of the public opposition relied on emphasizing the scientific authority of its own publications in implied contrast. Opposition materials sent to the press and public often listed at great length the credentials of the authors or the corroborating scientific materials—most often the two published parts of the NAS-NRC report (listed as two separate papers)—or included footnotes referring to those materials. *Fact and Fancy* included three footnotes, two citing the NAS-NRC reports and one quoting Decker. McLean's lengthy *Necessity, Value, and Safety of Pesticides* was peppered with citations of case law. Starc's *Nutrition Reviews* tract carried seven footnotes, of which four referred to other trade journals, one cited one of the NAS-NRC papers, one was a self-reference, and one referred to a general science text. Although the intent of these ploys was clearly to match or outgun Carson's science, their effectiveness is difficult to evaluate at this point. But compared with Carson's fifty-five pages once the book was available, a handful of footnotes cannot have seemed so very impressive.

Ultimately, however, Carson's opponents were called on to answer the very powerful but apparently "lone" voice of a citizen-critic, which they tried to do using almost every means other than writing a book themselves. The reasons members of the opposition did *not* write their own book probably encompassed many considerations, including the time required to write and produce a book that might or might not be a best-seller, as well as the need to find a comparably popular author. Most likely, however, the decision was based on the presumption that their own efforts to reach the public would be sufficient, even more so if the audience could be dissuaded from reading the book in the first place.

Reaching the Public, Using the Media

Opposition strategies to reach the public could be—albeit artificially—divided into three tasks: first, to appeal to the public audience itself, both directly and via the media as pass-through agent; second, to address efforts directly to the media, engaging them as partners in the debate; and third, to manipulate the media context in which the first two tasks took place, which could include efforts to keep the public from reading the book at all. The vectors of their efforts were often identical to those of author and publisher.

Particularly for commerce and politics, a basic form of direct communication to the public is advertising. In archived material on the *Silent Spring* debate, there were occasional references to ads placed by the companies and trade associations touting the social benefits of pesticides and addressing the issues of safety. These ads, however, were rare in the mainstream media. For one thing, issue-oriented advertising was only beginning to be common practice at the time, and the cost was usually borne by an individual company. Thus it is possible that other means of reaching the public were thought to be more cost-effective, notably mass mailings and public appearances supported by the trade associations.

Address lists for the mass mailings included important private citizens as well as members of Congress and, presumably, government agencies not already in sympathy with the opposition's views. Packets also were forwarded to Houghton Mifflin and Carson from recipients within the general population who had received them by dint of personal interest, membership in interest groups such as garden clubs, or stature in local government or society. Direct-mail address-list services were beginning to multiply in those years; and trade and agricultural groups maintained effective networks among themselves, augmented evidently by contacts in the medical fields—the American Medical Association, among others—through the efforts of Stare, White-Stevens, and like-minded others. One report of Stare's individual efforts came to the attention of Houghton Mifflin Vice President Lovell Thompson in a letter from an antagonized neighbor, who wished to inform Thompson that Stare was sending reprints of his article around to influential doctors.[88] More generally, *PR News* reported that ultimately "more than 25,000 reprints and 10,000 extra copies [of the Monsanto parody] were distributed" and claimed that "requests came from government officials, educators, businessmen, and Monsanto's customers, prospects, and stockholders."[89] Meanwhile, *Printers' Ink* noted that "quarterly mailings on the facts behind pesticide use are being sent to opinion leaders in science, industry, and government. One of these is [NACA's] *Fact and Fancy*."[90] Later, the *Wall Street Journal* reported that the number of individual addressees had topped 100,000.[91] Direct contact was made through means other than mailings, as well. "I have just heard that one of the chemical companies is now inserting its propaganda in the supermarket publications," Brooks wrote to Carson in November of 1962.[92]

Some of the material provided by the trade groups was expressly designed to help local agents or retailers deal with expected questions from the populace. In many cases, local workers or authorities such as state agricultural agencies saw their own practices and policies as being under attack. Responding defensively or offensively, with rebuttal or education, was thus important to their professions or even their livelihoods. One pest control trade organization offered guidelines: "You should have no reticence about assuring your customers or the general public that your use of pesticides . . . represents safe practice measured by the best scientific knowledge of today. Reaction to these articles will be quite varied. Yours, we hope, will be one of objective evaluation."[93] The *County Agent and Vo-Ag Teacher*, circulated widely by the NACA, explained "How to Answer Rachel Carson—to counteract anti-pesticide propaganda."[94] In many such articles, the address of the NACA or the Nutrition Foundation was provided, with an invitation to write for more information or to give copies of the articles to customers.

Meanwhile, direct personal contact was seen as a valuable form of communication, so long as it took place in public. Individual lectures and participation in public debates took representatives—notably White-Stevens—far into the communities. White-Stevens's primary occupation during the episode was to make such appearances all over the country, but Stare, Baldwin, and more local "pesticide authorities" also went on lecture circuits. These events went well beyond presentation at specialized professional meetings to such lay-audience venues as club meetings and community forums in schools and churches. Government representatives (notably ARS's Byron Shaw) proceeded in much the same way as industry spokesmen. They were often called on to speak for the USDA before the public and the media—albeit carefully, with guidelines firmly in place. Fifth on an official list of five actions to be taken in conjunction with circulation of the USDA policy statement on pesticides was the following: "Utilize every opportunity in forthcoming speeches, and in our information activities generally, to tell what the Department is doing to safeguard the Nation's food supply."[95]

Parallels between this grassroots dissemination of the opposition's message and efforts on behalf of the book—mailing lists, contacts with conservation societies and garden groups—are inescapable. Carson's personal appearances, though limited by health or contractual restrictions, often involved similarly local or limited groups. Likewise, the speech circuit of chemical company representatives mirrored to a certain degree appearances by Audubon Society members and others committed to "the cause," including Ford and Brooks. Despite the amount of resources devoted to such immediate, largely unmediated contact with the public, however, it was the reach of the book *itself* that worried the opposition. *C&EN* emphasized concern that a large audience could be expected, given the two million copies published of Carson's previous bestseller, *The Sea Around Us.*[96] Even with mass mailings, direct contact with that many people was a daunting task, but the compounding media attention meant that Carson's opposition could hardly stop at direct public contact.

At a pivotal threshold of that media attention lay the entire realm of book reviewing. Well-placed reviews could allow articulation of the opposition's most basic disputes with *Silent Spring*, assuming that the reviews were written either by spokesmen for the opposition (such as Darby or Baldwin) or by credentialed sympathizers, especially those who might be quoted in handouts or letters to editors. With few exceptions, however, rebuttal-type reviews appeared largely in trade and professional magazines rather than in mainstream magazines and newspapers. The two most visible exceptions to that rule were *Time*'s harsh review, which included a quote from Decker, and the *Saturday Evening Post*'s scathing commentary by Edwin Diamond, the journalist once hired and then dismissed as Carson's collaborator. Almost as helpful were book reviewing organs that allowed for more than one comment on a controversial subject. A good example was the *New York Times Book Review*'s offering of contrasting comments by NACA executive Parke Brinkley in a sidebar to its review of *Silent Spring.*[97] Another attractive aspect of book reviewing was the implicit invitation to write a letter to the editor, an invitation Carson's critics often accepted.

Otherwise, the opposition dedicated most of its efforts to gaining the media's attention and, where possible, affecting the media's treatment of *Silent Spring.* The mass mailings were, after all, aimed substantially at the media. The press received policy statements from government and press releases from industry and trade groups, often accompanied by some assortment of enclosures from among the eight core publications, especially "Desolate Year" and *Fact and Fancy*. According to *PR News*, five thousand copies of "Desolate Year" in galley-proof form had gone promptly to "book reviewers; syndicated science and gardening writers; editors of Sunday, farm, and science papers; trade publications in eleven big industries; national magazines; and radio and TV farm editors."[98] Several packets carrying the opposition's reprints and articles in various combinations came from the NACA, the MCA, the Nutrition Foundation, and related groups. *Printers' Ink* reported, "MCA . . . has been mailing news feature material on the benefits of chemical pesticides to 2,200 publications a month.

Radio stations have been receiving similar feature items, one of which was titled 'You Almost Didn't Get Your Orange Juice This Morning.' "[99] The MCA was reported to have sent a derisive cartoon to 50 dailies and 800 weeklies.[100]

USDA interactions with the media also went beyond issuing official position papers and scripted debates, even if the transactions were not always public. In correspondence between E. G. Moore of the USDA and Harland Manchester of *Reader's Digest* (which had declined to publish any part of *Silent Spring*), Moore followed up a USDA meeting with Manchester by commending to him the negative reviews from *Time* magazine and from Darby, which he enclosed, calling Darby "one of the best qualified persons in the United States to comment on *Silent Spring*."[101] Similarly, midwestern columnist R. Milton Carleton, who was also head of a Chicago commercial "garden research center" and ardent critic of Carson, drew on the department's resources to arm himself for a possible debate with her (which never took place). Moore told him, "You are taking on a tough assignment if you agreed to debate with Rachel Carson on television. People are guided pretty much by emotions and, of course, she will have you licked on that score."[102] Considerable correspondence between Jukes and the department included Jukes's own recommendations on managing media coverage. Links between the USDA and more senior industry leaders meant that similar cross-consultations must have taken place as the controversy continued. But it was industry and not the USDA that had both mandate and resources to cope with the problems posed by *Silent Spring*, free to approach the media with whatever it chose to do.

Indeed the campaign took on such proportions that editors and book reviewers began to remark on it in print, and in the opinion of some, it thereby backfired. Following a conversation with *Boston Globe* book editor Ted Laycock, Ford reported, "The chemical people have been sending him oodles of stuff. He thinks it's outrageous, and doesn't agree with it. He says they will hang themselves, but he has to turn it all over to the powers that be."[103] More skillful ways to use the media to publicize the opposition's side lay in more indirect approaches.

Using and even creating newsworthy events that journalists are compelled to report is a time-honored publicity ploy. Had Carson's industry-based opponents failed to use that tactic, their public relations officers would have failed in their jobs. Accordingly, whenever White-Stevens, Stare, Decker, or any of the designated spokesmen-authorities for the pesticide industry spoke publicly, they made sure the appearance was reported as widely as possible and always mentioned their criticisms of *Silent Spring*. A meeting of the American Chemical Society in early September, for example, was used by USDA's Thomas Harris as well as NACA's leaders as a national—and highly credentialed—platform from which to enunciate their criticisms.[104] Similar media-magnet debates were arranged around the country, often on local radio, even though Carson routinely turned down invitations to attend. Those debates not only constituted their own form of exposure but also often prompted coverage by a story-hungry community press.

Frequently, one part of the media system was drafted to cover another: newspapers covered radio debates and radio programs noted print book reviews; text printed in one organ—often the *New York Times* as iconic authority—was cited to amplify discussions elsewhere. Though the *Times* was an early, frequent, and emphatic voice supporting Carson and her alarms, Walter Sullivan's review in the daily edition carried both praise and criticism. His most negative comments accused her of one-sidedness, and he said she had opened herself to parody: "Some unsung hero of the chemical industry has written, for Monsanto Magazine, an article titled 'The Desolate Year.' "[105] That quote appeared widely in opposition material, with the *Times* always noted as the source.

Even though publicists for the opposition may not have been able to control exactly what the press said, these methods assured that they had enlarged their audience for the message that there *was* another side of the story. Whether Carson's opponents trusted or relied on journalistic objectivity, however, is open to question. Their behind-the-scenes legal efforts to stop or delay publication had a public counterpart in activities designed to persuade the media to deny *Silent Spring* credence or, if possible, exposure in the first place. To counteract glowing reports of popular support for Carson, individual and group letter-writing campaigns from the opposing camp were soon organized. Their targets included not only local and national government officials and editorial pages in local papers but the management of the media itself.

As early as the end of August 1962, the word got out that CBS planned a documentary on the pesticide debate,[106] and Carson's opposition understood that meant exposure and perhaps validation at the broadest national level. In October the editor of *American Fruit Grower* told readers about the planned program and urged them to write the producers (at the time, named as Jay McMullen and Palmer Williams), giving them CBS's address, and advising them to "tell these gentlemen that you expect a fair and impartial discussion" that would focus on the benefits and safety of pesticides in food production.[107] Other publications made similar suggestions. CBS's Fred Friendly commented to the press on the advance mail at the time, and McMullen later confirmed that a letter-writing campaign had taken place: "We got more than a thousand letters, which was an astonishing amount of mail—never had that much mail before a broadcast went on the air. First of all, they're all mimeographed; they all said the same thing, in essence, 'Look, be fair about this issue.' I thought they were from the chemical industry, but I could be dead wrong about that."[108]

CBS would soon feel more serious and high-profile pressure from its advertisers, although the actual circumstances of the episode had more than one interpretation. Shortly before the April 3 *CBS Reports* program was to air, the press announced the withdrawal of some of the advertising for the broadcast; the defectors were food producers Standard Brands and Ralston Purina, and Lehn and Fink Products, makers of Lysol antiseptic spray, leaving only Kiwi Polish and Brillo as advertisers for the evening.[109] The official position taken by both CBS and the withdrawing advertisers was that the ads had not been canceled, merely shifted to other time slots[110] or later *CBS Reports* segments for what amounted to aesthetic reasons and not because the companies objected to the content of the program. Lehn and Fink claimed to have "no opinion about the nature of the program; our marketing people just decided that the show's audience wouldn't be the right one for our products."[111] One trade publication reported that Lysol just "wasn't exactly the type of product to be advertised on a show about insecticides." But according to *Broadcasting* and Ralston Purina spokesmen, CBS had alerted the advertisers to potential incompatibilities between the show's topic and their products, giving the companies the option to pull or switch their ads.[112] From CBS's point of view, the reduction of ads had the positive effect of creating extra air time for the program. There is no evidence that the three firms that pulled their advertising had any relevant connection with members of the NACA or the MCA, although conceivably the food producers could have had some involvement with the Nutrition Foundation. What is important, however, is the fact that any withdrawal of advertising was itself newsworthy, particularly in conjunction with the letter-writing campaign; and Carson's critics used that news as testament to the strength and seriousness of opposition to her message.

Books and Debate

Hindsight is always clearer than foresight, so it is easy to infer that engaging *Silent Spring* on so large a scale gave the book and its message considerably more attention than they would have received had Carson's opponents exercised more restraint. That inference may seem simplistic and obvious, but examining the "why" of it is instructive in thinking about how books have functioned in the American media system. First, at the most literal and basic level, all the efforts to match the snowballing reach of a bestseller—beginning with the "many other pieces that will be written as the result of it"—effectively constituted at the very least a doubling of publicity, if not magnification well beyond that. Even though the opposition's messages sought to negate Carson's warnings, the fact that such warnings existed in the first place was inevitably part of the story. Second, and almost as basic, efforts to both undermine and match the inherent authority of the book form actually underscored that very authority. An important part of both processes was that rebuttals by Carson's critics prompted Carson's supporters to rebut those arguments. Challenges to Carson's intrinsic message and credibility gave Houghton Mifflin, as well as advocates and supporters in government, conservation, and politics, the opportunity to reassert the validity of author and work regularly and in an ever-expanding forum.

Quite simply, having framed *Silent Spring* from the beginning as a gratuitous, unilateral attack, the counter-Carson campaign created a debate and ensured its continuation. That debate became, further, a very public one, conducted in the public forum of the media system at all levels, from very local to national (and eventually the international). Moreover, in framing the issue of pesticide use and policy as a debate in the public forum, Carson's critics created an entity more attractive to the media than the simple fact of a book's publication. Journalism loves controversy, not just for the excitement but also because it promises at least two sides to any issue, and critics of *Silent Spring* provided, ready-made, all the material for that second (and third and maybe a fourth) side of what might otherwise have been a dry, technical topic. More essentially, it is a truism of journalism that *events* are more easily covered than *issues*; and a debate— one with confrontations, meetings, lectures, shifts in position, revelations, charges and countercharges—provides a continuing series of newsworthy events.

Although Rachel Carson was clearly their designated opponent, her challengers' view of who was the audience and who were participants in the debate is less clear. The public forum into which the opposition brought the debate encompassed the level of individual citizen but seems to have been focused on the media and their interlocking influences rather than on that single reader. A retrospective comment by Jukes conveys the sense that the general public was originally not considered a participant in the debate but rather an observer: "There is a great value in bringing the public into these debates, but the public have to be supplied with authentic information, and we felt that *Silent Spring* had not done that."[113] The logical inference is that prior to *Silent Spring,* debate on pesticides had existed but existed mainly outside the public sphere. Carson's assault had required and justified a fully public argument, best carried on through the media. Thus, where author Carson thought of her final targets mainly in terms of the collectivity of activist reader-citizens, her opposition's energies—perhaps because they were designed to fend off action rather than mobilize it—were devoted to manipulation of the media. Though each side had an eye to government practice and policy changes, the opposition's hope was that policymakers would necessarily interpret the flood of information to the media as proof of popular rejection of *Silent Spring*'s warnings. Further, with enough airing of the matter in venues where the opposition had some control, the general public—by this reasoning— might be convinced that reading the book would be unnecessary or, better, inadvisable. That reasoning underestimated the power of the spotlight the critics themselves had thrown on *Silent Spring*, and it underestimated the power of a message conveyed in book form, even if *not* read.

CHAPTER 5

Media: "One Formidable Indictment"

Given how clearly Carson and her publishers relied on the media to extend the reach of the book, the impulse may therefore be to think first about what the media contributed to the book's public life. But that discussion risks severe superficiality if we fail to understand first what the media received from *Silent Spring* as a book carrying information and warning. Only after examining what *Silent Spring* "did" for the media is it possible to understand fully the media's contribution to the life of the book, which far exceeded the simple spreading of word about the book's existence.

What *Silent Spring* Did for the Media

Publication of *Silent Spring* and Carson's critics' countercampaign had together transformed the matter of pesticide use—and the larger matter of human intervention with nature—from a difficult-to-cover, complex *issue* to one newsworthy *event* that spawned other newsworthy events. Although the problem of pesticide abuse was not entirely new to the media, pesticide abuse had never been seen as a particularly newsworthy topic per se; and photos of dead wildlife had had limited ability to arouse the general public. Moreover, the press lacked the research resources and—for the most part—the editorial skill and leisure to explain clearly and compellingly what was happening and why the public should care. Even if a publisher or broadcaster had the will to try, the almost universal expectation was that only a sensational occurrence could capture public attention enough to justify covering so difficult an issue. Few would have predicted, however, that the sensational occurrence would be publication of a book.

When *Silent Spring* emerged, it provided the media with an organized, rationalized collation of all the arguments and supporting material for the case against pesticide misuse, in accessible language that easily related to individual readers' interests. As *Newsweek* wrote, "Now all these arguments against pesticides have been assembled in one formidable indictment."[1] In one place, readers, reporters, and editors could find understandable explanations of scientific processes, criticism of established practice and policy, the supporting evidence to verify the charges, and a vivid, immediate description of the deplorable consequences.

More than a convenient shortcut for editors and reporters, however, *Silent Spring* brought the weight of the book form. That weight included the inherent weight of the author's work and reputation, the added weight of the institutional support of an established publisher (Houghton Mifflin), and the almost literal weight of its fifty-five pages of references. In addition, appearance in the *New Yorker* had added its own institutional status to the cultural "heft" of the book. To torture Bourdieu's concept of cultural capital somewhat,[2] the book had a function that could be described as a sort of "intellectual capital," underwriting the media discussion of *Silent Spring* and the attendant issues in the public sphere. The media could feel confident about the validity of media-borne debate on pesticide abuse because it revolved around a book produced by a nature writer of some repute, grounded in visibly extensive research, and presumably vetted and supported by two highly respected publishers. Reporters and editors could credibly refer to the complexities of pesticide policy and practice without actually having to pursue the intricacies themselves—in fact, without having to do more research than reading *Silent Spring* and perhaps some of the countering material supplied to them. Similarly, insofar as Carson's critics challenged her and presented their own arguments in ways that mimicked the book form—with scientific references, credentials, and footnotes—they too sought to be underwriters of the general worth of debate negotiated in the public forum.

In the media's own collective consciousness, however, the worth of *Silent Spring* was not understood in such theoretical or even processual terms. For them, the initial question was that of newsworthiness; and by itself, book publication did not constitute an event remarkable enough to sustain much media attention, although some tried to make that claim. The book's news value was commonly invoked by declaring the book "important" with assertions about just how important it was. U.S. Supreme Court Justice William O. Douglas's pronouncement that it was "the most important chronicle of this century for the human race" frequently appeared in the general coverage. Others, like the Omaha *World-Herald*, would echo the theme: "one of the most important books published in our time."[3] And headlines like the Cincinnati *Enquirer*'s "Could Be Year's Most Noted Book"[4] were common.

Why the book was important was not always explained, but the media often remarked on the possible impact it might have on public perception and action. As attention mounted, so did the estimations of its effects: "It will go down as one of the intellectual blockbusters which significantly influence human behavior," wrote one reviewer.[5] The idea that, indeed, informing and mobilizing were evidently Carson's mission placed the book squarely within journalism's traditional image of its own mission. At the very least, the media described her work in terms of having performed a public service, implicitly much like their own in reporting on the book.

By the very fact of its advocacy, *Silent Spring* presented the media with the prospect of an ongoing story that was both larger and journalistically more attractive than pesticidal hazards—namely, a public debate. Indeed, in the short period after the first *New Yorker* appearance but before the controversy fully developed, some editors felt compelled to respond to Carson's message by suggesting that a debate *should* take place—that there might be another side whose views should be brought to light. Harry Nelson of the *Los Angeles Times*, for example, decided to report on the Monsanto parody because "the general public will not have access to it, the opposite side of the coin."[6] Certainly, the media's ostensible duty to practice objectivity and fairness would lead them to search out responses to Carson, but had *Silent Spring* itself been a detached examination of all the possible views on pesticide use, there would have been little media excitement about its publication. What especially excited the media about the *Silent Spring* debate were the personalities and events that enacted the debate before the public eye. Each time a positive review of *Silent Spring* was countered by a corporate spokesman, each time a lecture was scheduled for rebuttal to Carson, each time a community panel was formed to consider both sides of the question—each such occasion was a reportable event.

A debate is, after all, a drama of conflict, with dramatis personae, themes, narrative, and plot, and the *Silent Spring* debate was rich in all of these. The narratives of confrontation used words like "hubbub" and "uproar"; and the "battle" between "man and bug" or "man versus nature" paralleled the "battle" between Miss Carson and industry spokesmen. "Miss Carson has expanded the theater of war," wrote a *St. Louis Post-Dispatch* commentator.[7] Sometimes the image of a legal battle brought courtroom terminology like "counter-charges," "rebuttal," and "defense." An Indianapolis reporter wrote, "She presents her evidence like a public prosecutor, with a relentless battery of testimony."[8] Even in less combative terms, headlines would announce the controversy—"Naturalist-Author's New Book Worries Chemical Industry"[9]—and the text would juxtapose paragraphs listing her charges against paragraphs listing the opposition's. First-instance reporting was likely to mean merely taking statements from each side: the opposition was "aghast,"[10] while Carson claimed she was being misrepresented. When those statements were made publicly in person, moreover, they provided the added newsworthiness of a political or social context. Such occasions were understood and expected by all— author, opposition, public-relations professionals, and the content-hungry media mill—to serve that purpose.

In actuality, Carson and her critics met face-to-face on almost no occasion. The closest juxtaposition was the *CBS Reports* parallel interviews. But their arguments met "face-to-face" in events well suited to press coverage, in venues from high school auditoriums to Capitol Hill. Replicating national debate, local experts or book reviewers offered their own discussion of one or both sides—and these were covered by local media. Community book review groups took up discussion not only of the book but also of its critics. Panels of local scientists or gardeners were assembled for PTA meetings and local radio shows, often with one side or the other "loaded" with a particular local expert, adding attractive local color for newspaper editors. A typical example was the *Hartford Times*'s coverage of a local entomological society meeting (an event otherwise very unlikely to command media attention) at which a state agricultural agent spoke: "In his prepared answer to the 'Stone Age' accusation, Mr. Turner retorts that the argument of Miss Carson savors of the Dinosaur Age. That is older than the Stone or Neanderthal Age."[11]

Carson herself was a classically appealing protagonist, despite her best efforts to remain private. Much of the news coverage began with a description of her appearance and various qualities: "shy," "petite," "soft-spoken," or less felicitously, "spinster" or "bachelor biologist."[12] Although Carson granted few interviews, interviews and profiles by Barbara Yuncker for the

TO REVIEW BOOK — The Book Review Panel of Oakmont senior students presented their program to the members of the Literature and Arts Department of the Woman"s Club yesterday and will speak again Friday, April 19 at their school assembly and Saturday at the D.A.R. meeting in Vandergrift. Members of the panel are: Linda Lascola, Ann Gilbert, Kathy Ross, Jack Sherwin, Paul Hlavac and Andy Sloan.

Photo of book club from Oakmont, Pennsylvania, newspaper. Copyright © Oakmont Advance Leader. Yale Collection of American Literature, Beinecke Rare Book and Manuscript Library.

New York Post[13] and Ann Cottrell Free for the North American Newspaper Alliance[14] were widely distributed, and her few appearances were well covered by the wire services. In regional and local newspapers, it was common practice to incorporate quotes without attribution from the handful of interviews or speeches into articles profiling Carson, giving the subtle impression that the reporter had had personal contact with her.

As Carson became less available, reporters bent on preserving two-sided coverage of the debate often lifted phrases or whole passages from *Silent*

Spring to fill in for Carson's side. Certainly, there was logic in doing so, in that it guaranteed accuracy in reporting Carson's arguments. But ethical considerations aside, something further was at work, as evidenced in the desire to have recordings of Carson reading her work for radio use and—above all—in Jay McMullen's heavy use of Carson reading from her book, visible to the camera, throughout the *CBS Reports* program. Using the author herself as part of the story was inevitable; ultimately, the result was almost full conflation of the author with her book and mission, such that "Rachel Carson" and "*Silent Spring*" became emblematic media shorthand for the entire matter of pesticidal hazards and, eventually, environmentalism itself.

Her critics, too, were rather memorable media figures in the debate's drama, particularly the authoritatively annoyed White-Stevens and the irascible Stare. Although the demeanor of such people was rarely described in the print media—in contrast to comments about "the diminutive Miss Carson"—their television personae were a striking part of the *CBS Reports* program. Similarly, radio addresses also showcased their rhetorical styles in a way that print could not. From local to national coverage, a battle of credentials and character between supporters and critics raged in subtext. And photographs often played out the confrontational themes—sometimes between scientific authority and "hysterical" nature lovers, sometimes between heavy-handed corporate mouthpieces and passionate but gentle conservationists.

Any drama of confrontation, however, craves resolution, with a winner and a loser. Without resolution, moreover, there was danger of losing public interest. In this, the media—journalism specifically—faced an acute and abiding quandary about the role of the press in public controversy. The traditional assumption has been that objectivity is the proper journalistic goal, and that fairness and truth are part of that objectivity.[15] What constitutes objectivity, and whether it is the same as balance, are time-honored philosophical questions that become practical matters when covering a public issue under debate. In the case of the *Silent Spring* debate, the opposition's charges of one-sidedness worked as an implicit demand of the media that *their* discussion of the issue not be similarly one-sided—in other words, that they not agree with "Miss Carson." Such demands worked to reinforce the press's own professional inclination to pursue fair and balanced coverage of an issue while at the same time working against the impulse to declare a winner and find the objective, or at least scientific, truth.

BUGS AND SPRAYS were the subject of scholarly discussion at the annual picnic of Wellesley-in-Westchester recently at the home of Mrs. Allan J. Newmark in Chappaqua. From left are, Mrs. Harold Pennington of Mount Kisco, picnic chairman, Dr. Cynthia Westcott of Croton-on-Hudson, Wellesley alumna and recognized garden authority, who defended the sensible use of insecticides, and Mrs. Edward S. Menapace of Eastchester, president of Wellesley-in-Westchester and hostess for the affair.—Staff Photo by Doris B. Kirchhoff.

County Wellesley Group Told 'Silent Spring' In Error

By SIDNEY EATON BOYLE

CHAPPAQUA—

The sensible use of insecticides was the subject of Dr. Cynthia Westcott, known as the "plant doctor," at the annual picnic of Wellesley-in-Westchester held recently at the home of Mrs. Allan J. Newmark, Laurel Lane.

Dr. Westcott, a Wellesley alumna, discussed Rachel Carson's book "Silent Spring," which warns against the use of pesticides.

The speaker, herself author of many standard reference and handbooks for home gardeners, disagreed with Miss Carson's frightening conclusions that people are being poisoned and our wildlife and vegetation are being extinguished by the use of chemical pesticides urged on us by groups more concerned with the industrial dollar than the common good.

Miss Carson especially objects to spraying programs which she believes have caused tremendous harm without any benefits.

Sees Little Harm

Dr. Westcott declared, "The gypsy moth has been cleared up in New Jersey and Pennsylvania with little harm even to bees—just a little loss of fish in unmarked ponds."

In New York, she said, the program has been less well-coordinated, "some areas being sprayed twice, some crops damaged, some fish killed, but when the case came to court the judge did not find the vast amount of damage claimed in 'Silent Spring.'"

"Having had a summer place on Long Island and living now in Westchester," Dr. Westcott said, "I can definitely testify that birds have not been silenced by the gypsy moth spray."

She said "Silent Spring" may well result not in decreased use of pesticides, but in more efficient and widespread use.

Sees Good in Book

The book can make casual or callous users aware of the potent qualities of these chemical tools, she declared.

She also expressed hope the book will spur accelerated research and development of improved versions and encourage understanding of the advantages as well as the limitations of the chemicals.

She concluded by warning users to keep the chemicals in their original containers out of the reach of children. They should not be stored in cabinets with food packages, she advised.

Prior to spring 1963, a good deal of the discussion of *Silent Spring* and its critics carried the implicit or explicit question "Who is right?" This question is a much easier one to report on and to debate than "What is the truth?" even if they are logically the same. Pairing Carson's charges—for example, that appallingly irresponsible accidents had occurred, with deadly results—with illustrative local incidents constituted submitting those events as supporting evidence to "find for the plaintiff." Similarly, taking statistics like White-Stevens's—that in one year only 89 people had died from pesticide-related poisonings, compared with 128 who had died from aspirin (presumably overdoses)[16]—and combining that data with industry statistics about the size of the fruit crop relative to the size of the need for fruit might show the weight of the "rational" arguments against anti-pesticide "hysteria."

Media coverage of *Silent Spring* thus oscillated within a four-way impulse to (1) see value in Carson's warnings, (2) find merit in criticisms of *Silent Spring*, (3) seek objective "fairness" in debate coverage, and (4) declare a reportable winner to complete the drama. Such tension is, ultimately, the tension between journalism's mission to inform and mobilize and the journalistic ideal of impartial objectivity. In theory, the tension is resolved through separation of detached reportage from editorial advocacy. In practice, that partition is often difficult to maintain, as was often evident in media treatment of *Silent Spring*. To avoid the appearance of appointing themselves judges, the media often resorted to a thorough airing of both sides, thereby making themselves not just sources of information but also catalysts of the controversy. In doing so, they had the added effect of prolonging the debate by replicating it. The *CBS Reports* press release for the April 1963 broadcast had set the stage for an impartial presentation "voicing a broad spectrum of opinion,"[17] and the program's format followed an almost mathematically balanced matrix of time slots allotted to the various spokesmen and Carson.

By spring of 1963, the continuing lack of resolution on the issue seemed to trouble the public or at least the media. When the initial *CBS Reports* program aired, some reviews of the program announced that Carson had been "the winner" or at least had "held her own."[18] A number of television critics (and letter writers), however, complained about the absence of conclusions, especially after all the preliminary hype (including the withdrawal of advertising).[19] The impulse to find some resolution to the controversy, to find out "who's right," ultimately led much of the media to treat the release of the PSAC report on May 15, 1963, as that resolution, casting the presidential committee as final judge and Carson the victor. The marked drop-off of mentions of the book thereafter strongly suggests a sense that coverage was allowed to ease because a winner had been declared and therefore the debate was over. The forthcoming congressional hearings were treated as little more than evidence of the win and of interest only to inveterate Washington-watchers. Although for Carson those hearings signaled just the beginning of the kind of political action she always had in mind, they were of far less interest to the media, perhaps, than its coverage of its own coverage.

Silent Spring had provided to the media, first and last, the insertion of a complex issue into the media's agenda. It did so as a single source of compiled and organized information ("one . . ."), a source that carried the weight of the book form, the publishers' imprimatur, and the author's intellectual work (". . . formidable . . ."), and one containing criticism and warning guaranteed to provoke public controversy (". . . indictment"). That "one formidable indictment" was something the media would have been hard-pressed to provide for themselves, despite intermittent efforts to imitate or to duplicate *Silent Spring* without direct reference to it. Carson's advocacy in writing *Silent Spring* and in getting it published was essential to its agenda-setting function; and in that sense, her mission meshed nicely with the journalistic mission shared by most of the media. Moreover, the constituents of the debate as well as the debate itself—both easily attached to an emblematic book, author, and opposition—supplied controversy in a form eminently suitable to the media's needs for personification, drama, and resolution.

What the Media Did for the Book

Clearly, it is simplistic to say merely that the media extended the reach of the book, which by itself would have reached only a comparatively small audience. We have already recognized that function, as did Houghton Mifflin's publicity department, Carson's agent, and Carson herself. The media were, however, far more than the vectors by which an expanded audience got word of publication. Word of the debate came soon after publication; and coverage of the *Silent Spring* debate was quantitatively, qualitatively, and chronologically more extensive than coverage of the book per se. The replication of the debate took place throughout an arena occupied by the media, public actions and interactions, and most important, the intersection of the two—a nexus to which Habermas's idea of a "public sphere" might arguably apply.[20] Moreover, in the processes of its replications, the debate took on a richness and texture that one can only begin to suggest, let alone discuss in an organized way.

Book reviewing would seem the obvious place to start, as a threshold through which announcement and evaluation of a book is disseminated to a new audience. A rich and significant "culture" of book reviewing has been identified, with distinctive cultural, social, and political features equally applicable to nonfiction and fiction, as well as to trade, scholarly, and mass market books alike.[21] With respect to nonfiction carrying a message like *Silent Spring's*, the existence of a politics of reviewing is particularly significant, involving such considerations as choice of book for review and even more, choice of reviewer. It might be logical, for example, or even required by the charter of a publication, to choose a specialist knowledgeable about a technical topic, such as chemical pesticides, to review a book about that area. But if the editorial policy of a magazine or newspaper so dictated, the specialist could be an archrival of the author of the book to be reviewed—as in effect happened with such reviewers of *Silent Spring* as William Darby of *Chemical and Engineering News* or I. L. Baldwin of *Science*. Moreover, particularly in scholarly circles, book reviewing can be both literary criticism and an exercise in philosophical debate. The responses of one reviewer may be noted or even questioned in yet another's review, sometimes identifying the first reviewer by name. Thus a debate among reviewers—sometimes with, but usually without, the participation of the original author—is carried out in public view.

In the case of *Silent Spring*, however, the controversy pervaded the media in such a way that differentiating reviews from general coverage and opinion was sometimes impossible. Author and reviewer Oliver LaFarge noted in an early column, "Here is a startling book that might well be handled in the news columns rather than the literary columns of any paper."[22] Often the distinction between news, press commentary, and book review was nothing more than the presence or absence of the standard parenthetical notation of publisher and price—"(Houghton Mifflin, $5)"—within the text of a formal review.

Admittedly, the virtual indistinguishability of book reviews from other coverage of *Silent Spring* is partly explained by the appearance of the *New Yorker* articles before publication, obliterating the book's debut as the starting point for the debate. Also relevant is the fact that many more book reviews appear in local newspapers than in literary magazines; and with nonfiction particularly, local readers may consider literary discussion secondary to judgment on the book's arguments—judgment that need not take place within the reviewing format, which is often more interested in form than content. Thus it was not uncommon to find a discussion of *Silent Spring*, perhaps a synopsis of its content and possibly a comment on its worth, presented not as a review but as a bylined article or column in the first section of a newspaper. Indeed, the news value of both book and debate conditioned editorial decisions about where to discuss *Silent Spring*. The *Houston Chronicle*'s Tom Mulvaney explained his own choice to Houghton Mifflin's Ford: "In my dual job on the *Chronicle* as editorial page writer and book editor, I had the choice of doing a review on the book or writing the editorial. I chose the latter in the hope the message would reach a greater number of readers."[23] For these reasons, isolating the part of the debate that was carried on within the book-review pages for separate analysis would result in an exceptionally distorted and limited view, implying a difference that did not exist. Above all, it would miss the crucial point that the debate was carried on in a media continuum that largely ignored differentiation of book reviews from editorials, columns, and general coverage.

The expansion of the book's audience into the full media system was, moreover, more than a simple arithmetical multiplication of the number of people made aware of the book and the debate. Nor were the media merely blank slates on which Carson and the opposition scratched their arguments. As the book gave the media a reportable controversy played out in reportable events, so the media saw merit in playing a continuing, catalyzing role in the debate itself. Newspapers and magazines were likely to adopt the basic question as to the validity of Carson's arguments and make it into the headline for their initial coverage of the book: "Weed Killers or Man Killers?"[24] "Are We Poisoning Ourselves?"[25] At least one newspaper made the invitation to argument explicit when the Monsanto parody "The Desolate Year" became available and the *Richmond News Leader* elected to reprint it, explaining to readers: "Our thought is to invite discussion and to urge the widest possible hearing for what she has to say."[26] Reporting on the flood of material from the opposition, moreover, served to guarantee that a debate would arise in an indirect sort of "let's you and them have a fight" ploy (not to mention providing a classic example of media's cardinal self-fulfilling prophecy that something receiving a lot of attention is sure to receive a lot of attention).

The media's role did not stop at providing a courtroom for public exchange between adversaries and encouraging confrontation. The media themselves became participants in it, by indirect influence as well as direct intervention. Attendant within each one of the thousands of clippings preserved in the Carson Papers archives[27] was a reporter's, columnist's, reviewer's, editor's, or letter writer's perspective. Editors decided what to cover and which letters to print, but they also wrote editorials about the issue. Reporters made decisions about how to describe Carson as well as how to depict squabbles at local garden clubs. Columnists made their pronouncements about whether Carson was hysterical or the pesticide advocates were heavy-handed. And once they moved from journalistic objectivity to advocacy, those in the media became overt partisans in the debate. Even without choosing sides, in speaking in their own voices, they expanded the breadth and depth of the debate and further embellished its terms.

Furthermore, thoughtful book reviewing, editorializing, and commentary almost require contribution of the writer's own experience, insights, and philosophy. Garden-page columnists might report their own pesticide practices; sports magazine writers might mention the disappearance of favorite fish. Many reviewers used the occasion of *Silent Spring* to expound on their own views of human endeavor with respect to nature; perhaps the epitome of that was Eric Sevareid's various published columns, radio broadcasts, and certainly his commentary for *CBS Reports*. More than half of his initial response to *Silent Spring* was devoted to pronouncements such as: "It is quite wrong for us to assume that in atomic war lies the only danger of 'setting back civilization' a thousand years. . . . The new religion of the scientist-philosopher, like the old time religion, invokes the sanction of hell-fire and damnation—but with proof."[28] Although few were as florid, assessments such as Sevareid's provided something far more than a simple pro or con judgment. They elaborated the core debate, developing it with new perspectives that, while they may have indicated support of one side or the other, took the discussion well beyond the initial framework provided by Carson and her public adversaries.

Each time the book was discussed in a personal article or special venue, moreover, the color of a locality or the concerns of a special interest were added to that corner of the debate. Although the book was always the focal point, the media supplied a custom-tailored context for each pocket of audience while embellishing new levels of replicating detail, fractal-like, into the fundamental debate. In geographic terms, coverage purposely made the book a local event, related to local issues such as the long history of Cleveland's battle with Dutch elm disease, or the interests of fruit growers in California, called "our agricultural area where the battle between man and insect over every crop is continuous."[29] The World Health Organization expressed worry about control of malaria and hunger. Parents expressed worry about protecting children from accidental contact with chemicals they had formerly not thought toxic. Farmers' wives expressed worry that their husbands' illnesses were related to pesticide exposure. The debate was ramified and potentiated as various subpopulations of the audience took it on, processed it in their own terms, and fed it back into the media machine through letters, meetings, press releases, or guest columns.

Finally, as if the picture were not already enormously complex, one abiding media habit created further intricacy almost impossible to untangle. Information flows freely within the media system, from one medium to another and from one organ to another, copyright laws and competitive practices notwithstanding. They borrow, buy, and steal from one another, argue with one another, and in fact depend on one another for new material to keep that media mill churning. The simplest formal examples are sales of wire-service stories and syndicated material generated by one organ and distributed contractually to others. A chronological survey of the newspaper coverage reveals that thanks to wire services and syndication, early mentions were just as likely to be found in Iowa as in New York or Washington, D.C. One of the earliest references to the *Silent Spring* magazine series was, in fact, was a front-page cartoon in the *Des Moines Register* in July 1962.

Furthermore, the media also reported on one another, reporting in one place that *Silent Spring* was creating considerable stir in another. A "Voice of the People" guest editorial in a Des Moines newspaper might be picked up and run in the *Chicago Tribune*. Further, one medium might often indulge in evaluating the performance of another in the conduct of the public debate. In late August 1962, ABC News had rebroadcast its "Focus on America" program about a crop duster named Al Lockwood and the dangers he faced. In her syndicated newspaper column, television critic Harriet van Horne remarked, "Considering the great hue and cry that has gone up lately about the contamination of our environment by DDT and such, it is a pity that ABC was not able to enrich this repeat film with an editorial footnote or two. Or, still better, a brief discussion by scientists setting forth the trouble we are borrowing when we make indiscriminate war upon the insect world."[30] Coverage of *Silent Spring* included a great deal of self-reference by the media, which ensured a continuing source of self-generated grist for the media mill; in the process the media rendered their status as full participants in the debate unmistakable.

Within the media system, however, the various organs and outlets per-

Cartoon from *Des Moines Register*, July 26, 1962. Copyright 1962, reprinted with permission of The Des Moines Register. Yale Collection of American Literature, Beinecke Rare Book and Manuscript Library.

formed slightly differing functions. While considering those differences, however, three conditions should be kept in mind: (1) the media never functioned independently of one another; (2) the adversaries in the debate never confined their participation to one or another genre; and (3) the audience did not receive information about the book and the debate solely from any one medium.

Newspapers

First, newspaper journalism provided the origins and vectors for much of the debate's spread outward from the *New Yorker* articles and the book— a diaspora that took place formally through syndication and wire service distribution and informally through cross-media perfusion. More particularly, newspapers functioned to bring both national and local focus to the debate with a sometimes intermittent rhythm but always with the ability to report on debate-related events in a timely and locality-specific way. Thus an ongoing historical record was created at the same time the dailies and weeklies were providing an immediacy and often human scale to the dramatized narrative of the debate. Further, they provided outlets for the personalization of editors', columnists', and readers' individual perspectives.

The greatest concentration of coverage was in the Northeast, for a number of reasons. Added to the location of Houghton Mifflin in Boston and the *New Yorker* in New York was the fact that book publishing was an important "hometown" industry in both cities. Carson was considered a hometown author by both Boston (because of the oceanic theme of her earlier books and because of her summer home in Maine, where Boston newspapers were widely read) and Washington, D.C. (except for her Maine summers, she lived in a Maryland suburb). And Washington was a company town for the federal government. The presence of a substantial upper middle class and upper class in the burgeoning Boston-Washington megalopolis seems also to have brought with them the wealth and leisure to foster popular love of nature in gardens and the great outdoors. In her two best-sellers about the sea, Carson had already appealed to those who spent time on the beaches of Maine, Cape Cod and the Islands, Long Island, the Delmarva shores, and the Outer Banks. And coverage of the Long Island aerial spraying lawsuit, which dramatized the issues in New York's own side yard, had already raised awareness and perhaps a disposition to outrage among New York area readers.

Of the New York newspapers, indeed of all newspapers, the *Times* coverage was by far the most extensive and usually the most supportive. Although the issue of pesticide abuse was not a new one for the *Times*, in 1962 its coverage on pesticides independent of stories related to *Silent Spring* was roughly four times that in the previous year. Because of the *Times*'s status as a national newspaper and an authoritative organ of record, its attention to *Silent Spring* and its controversy was guaranteed to be picked up by other newspapers around the country (as well as internationally). Awareness of the *Times*'s appraisal that *Silent Spring* was a pivotal event, if not a cause, was evident throughout media coverage.

The *Times* did not, however, adopt wholesale advocacy for Carson's side. Although its editorials, many articles by John Lee, and Lorus and Margery Milne's review in the *New York Times Book Review*[31] manifested a generally pro-Carson orientation, Walter Sullivan's "Books of the Times" column in the September 27 daily edition faulted Carson, saying, "In her new book she tries to scare the living daylights out of us and, in large measure, succeeds. . . . By stating her case so one-sidedly, she forfeits persuasiveness among those who know she is not telling the whole story."[32] Sullivan's ultimate assessment was positive overall, saying that like *Uncle Tom's Cabin*, the book was important *because* it was unfair and because it was timely and raised legitimate questions. Nonetheless, his negative comments were reproduced many times over by Carson's critics to bring the weight and credibility of the *Times* to bear on their own side.

Even more unfriendly, though subtly so, was what purported to be a summary of the debate in the Sunday *New York Times Magazine* run shortly after the *CBS Reports* broadcast in April 1963. Although Lawrence Galton began with a nod to Carson for having "delighted and encouraged" nature lovers and conservationists by "finally focusing public attention on the 'problem'," his analysis drew heavily and unquestioningly on material supplied in opposition mailings and the controversial NAS-NRC report. He quoted many of the foremost opposition spokesmen, including Stare, Baldwin, and Decker, and paraphrased industry assertions without attribution or investigation—and all without direct reference to *Silent Spring*.[33]

If the *Times*'s intermittent ambivalence might have given comfort or copy to Carson's critics, the *New York Herald-Tribune* added a bit of popularist color that appealed to critics and supporters alike, adding personal stories and occasionally inelegant language. It reported a flat-handed comment by the antagonistic Stare, to the effect that "the cranberry scare of 1959 was 'baloney' and the current pesticide scare looks like more of the same."[34] The word "baloney," which had so irked Carson and Rodell, recurred in headlines and commentary for the duration of the debate. On the other side, the *Herald-Tribune* outdoors columnist Art Smith twitted the pompous White-Stevens, calling him "the four-eyed, double-dome of the insecticide dodge."[35] Meanwhile, the Long Island press (*Jamaica, Long Island, Press*; Garden City's *Newsday*) cranked out copious articles on the debate in the context of the area's own experience with aerial spraying.

Elsewhere, the *Boston Globe* had already demonstrated interest in the issue, having published a 1958 series on the dangers of spraying campaigns. Its early August 1962 editorial applauding the public service of the *New Yorker* articles made a point of mentioning that Massachusetts already had legislation in the pipeline, by implication owing to the *Globe's* efforts prior to *Silent Spring*. When Carson spoke to the New England Wildflower Preservation Society in January of 1963, both major Boston newspapers were happy to report on it. The *Herald* included pesticide-related anecdotes drawn from areas as different as Framingham (Massachusetts) and a town in Turkey.[36] Boston was, however, also Stare's home base; and the *Globe's* coverage of Carson's speech was quite eclipsed by its seven-part series "In Defense of Pesticides," with Stare and the USDA's Byron Shaw presented as guest columnists (although Shaw's piece was actually a reprint of his *U.S. News & World Report* interview). *Globe* book editor Ted Laycock, in the process of letting Houghton Mifflin's Anne Ford know about the barrage of opposition material that he'd received, indicated that he'd regretfully had to turn it over to "the powers that be, and apparently they finally decided to use the latest stuff Dr. Stare did up."[37]

Washington, D.C., is both the seat of government and the geopolitical center for many national interest groups, such as the National Audubon Society, the National Parks Association, the Federation of Homemakers, and the Women's National Press Club—groups to which Carson spoke. Moreover, Carson had easy contact with notables who themselves were likely to command or to affect coverage locally if not nationally, not to mention her social connection to the owner of the *Washington Post*. In addition, Irston Barnes, chairman of the Audubon Naturalist Society and a friend of Carson, wrote a column for the *Post* with unwavering advocacy of Carson's point of view. But most coverage of the pesticide issue concentrated not on the rightness or wrongness of either side but on the governmental machinations and political consequences, reaching a climax with the May 1963 congressional hearings.

In comparing northeastern coverage with that of the Far West, South, and Midwest, a reasonable expectation would be that agricultural interests away from the urbanized Northeast would react with antagonism to a woman thought to seek a ban of pesticides, especially given the influence of the chemical and agribusiness industries. To a limited extent, largely in the politically conservative agricultural areas of California, that expectation is borne out. But the *Los Angeles Times* carried Eric Sevareid's rhapsodic musings on the significance of Carson's work as well as a considerable amount of coverage of the debate in the media and in government circles. Meanwhile, the tobacco states had recently been confronted with findings of arsenic in their products, which Carson had described in *Silent Spring* as "a classic case of the virtually permanent poisoning of the soil."[38] In the Midwest, sensitivity about the toxicity of pesticides predated publication of *Silent Spring*, and newspapers like the *Des Moines Register* and the *Omaha World-Herald* had readers and letter writers with personal experience of severe physical damage to people and animals from pesticide accidents. Like eastern urban areas, Chicago was concerned with garden and business interests as well as embroiled in its own controversy over spraying programs to control mosquitoes and Dutch elm disease. The *Tribune's* Howard James noted that the battle pitting tree lovers against bird lovers was one of long standing: "The controversy has rocked village halls, broken off friendships, and filled the air with emotional arguments from both sides."[39]

Thus any expectation of an East versus West or an urban versus rural polarity in the media's coverage of the debate was not borne out, although there was decidedly less attention away from the northeastern urban centers. As a rule, outside of the Northeast *Silent Spring* appeared on few book review pages; and much of the coverage of the debate appeared in syndicated stories and profiles, often cut to much shorter length than the originals. At the same time, away from the Northeast, the degree to which media treatment of *Silent Spring* adapted and tailored the debate to local concerns was, perhaps by necessity, greater than in media coverage in the Northeast, which was, by contrast, more interested in the political drama roiling about *Silent Spring*.

Magazines

Magazines take longer to produce and publish than newspapers; thus timeliness is generally less a priority than fitting a story into the magazine's framework and philosophy. Magazines, therefore, can usually afford to be more concerned with the ideas involved in an issue than with specific events attached to it. This freedom from time-critical coverage (news magazines excepted) explains in part why most magazine discussion of *Silent Spring* appeared after—sometimes long after—publication of the articles or the book. In addition, in many cases, magazine editors had likely become aware of the existence of the controversy through their own reading of newspapers. Finally, unlike the ongoing nature of newspaper reporting, magazine coverage is often a matter of spotlighting a matter a single time in one issue and then—with the important exception of letters to the editor—moving on.

Some magazines chose to approach *Silent Spring* as a matter of public education and information about pesticide use, often tailoring discussion to spring or summer editions. Others chose to confine coverage to the classic book review form, and even then many were not tied to the timing of the book's publication; some reviews appeared months after the controversy had cooled. The decisions of still others may have been affected by considerations of space or public relations or advertiser pressure from either or both sides. In any case, virtually all major magazines were compelled to respond to the debate in one form or another, but they did so on their own timetable. Few, if any, were detached observers, even if they strove for the appearance of impartiality; most felt entitled, indeed obligated, to present their evaluation to their readership.

For their part, the national newsmagazines were inclined to reduce the debate to its simplest terms: Carson versus the chemical companies, chemicals versus bugs. Replacing the nuances of local or special interests was the populist flavor of black hats versus white hats. By the very fact of the newsmagazines' national circulation, the appearance of the *Silent Spring* debate in their pages catapulted it into the uniformity of a fully national forum. Short of outright sensationalism, a sharpened sense of drama and conflict emerged, perhaps to enliven the potentially off-putting topic, perhaps to provoke action of one sort or another. The fact of weekly or even monthly publication, moreover, meant that newsmagazine editors had at best one or maybe two opportunities to convey the controversy and to provide their own perspective. Despite journalistic ideals of objectivity, partisanship was usually obvious, particularly in the case of *Time* magazine.

Prior to 1962, *Time*'s history with Carson had not been antagonistic, and in fact it had given her an award in 1952 for *The Sea Around Us*. Nor had it been hesitant to acknowledge the hazards of pesticides: in a 1949 article *Time* had called DDT worse than the insects it killed.[40] But between that time and its resoundingly negative response to *Silent Spring*, it had covered the 1959 cranberry scare in its "Bureaucracy" section, with almost unconcealed contempt for the government's decision to ban the cranberries. *Time* also had a history of enmity toward the *New Yorker* owing to a long-running feud between *New Yorker* founder Harold Ross and Time-Life's Henry Luce,[41] which might well have set the stage for a slap at Carson and thereby at the *New Yorker*. *Time*'s report on *Silent Spring*, "Pesticides: The Price for Progress," ran in the issue dated the same week as the book's publication and sported a photo of a spray plane fighting a "Biblical plague," with a portrait of Carson by Alfred Eisenstaedt (one also used later for the *Life* profile). The article ran in the "Biology" section (thus not a review) and opened with Carson's parable, complaining of her "emotion-fanning" language. *Silent Spring*, it said, was "a real shocker," not just to an "unwary" readership but also to "scientists, physicians, and other technically informed people" who would find it "unfair, one-sided, and hysterically overemphatic," with "over-simplifications and downright errors." Most of the article was devoted to refuting her charges of harm—dismissing reports of wholesale loss of bird life, asserting that only a relatively small proportion of the American land mass had been treated with pesticides, and offering supposedly reassuring reports of human guinea pig volunteers fed two hundred times the "normal" amount of DDT without ill effects within one year.[42]

U.S. News & World Report (USN&WR) had a comparatively small circulation (less than half *Time*'s[43]), with a reputation for exceptional access to government offices and somewhat conservative politics for the Kennedy era. It delayed taking any notice of *Silent Spring* until late November 1962, when it ran its long interview with Byron Shaw, administrative head of the Agricultural Research Service. The question-and-answer format of the interview permitted carefully designed questions and carefully worded answers, in language undoubtedly worked out painstakingly within the USDA beforehand. The article concerned the necessity of pesticides ("without pesticides, food prices would be higher"), the dangers posed by "foreign insects," the safeguards already in place, and the need for much more research. Carson's arguments were presented as four bulleted summary points within a boxed, three-quarters-column insert, concluding with a direct quote of her disclaimer that "it is not my contention that chemical insecticides must never be used."[44] *USN&WR* offered no further report on the debate until the PSAC report released in spring of 1963.

The magazine's first, brief mention of the report in late May said that not until that spring had the White House entered the debate "touched off by" *Silent Spring*—an odd misstatement considering Washingtonian awareness of the committee's work for the previous eight months.[45] The following week, a somewhat longer article titled "If You Didn't Have Poison Sprays" claimed that the benefits of pesticides had been "almost overlooked in the continuing dispute"[46] (another odd claim, implying that the magazine's own interview with Shaw had gone unnoticed). The ten-paragraph discussion of those benefits included a single sentence about the PSAC report, noting only that it warned "that use of insecticides involves possible health hazards that should be eliminated."[47] While less overtly hostile to *Silent Spring* than *Time*, *USN&WR*'s form of journalistic balance gave the sense that, as many in the debate felt, one side (Carson's) had already been covered to excess and the other side needed shoring up.

With a circulation only slightly larger than *USN&WR*'s, *Newsweek* had already addressed pesticidal hazards on several occasions. The magazine had covered DDT and pesticides in the latter fifties and early sixties, including the Long Island spraying trial, and in 1961 reported on severe psychiatric symptoms suffered by Australian greenhouse workers handling insecticides.[48] Its approach to *Silent Spring* was more as continuing attention to a known problem than as announcement of a new controversy. First notice in early August 1962 was a piece in the "Science" section that drew heavily on quotations from the *New Yorker* articles. Although the article stopped short of explicitly taking Carson's side, it did cite the *New York Times* editorial suggestion that she might be worthy of a Nobel prize.[49] A year later on release of the PSAC report, *Newsweek* conferred the status of arbiter on the government, its banner headline declaring "Judgment on Pesticides" in the "Medicine" section. But *Silent Spring* was the focus of the first half of the article, clearly implying that the government had been "goaded to look into the entire question" because of *Silent Spring* and responses to it.[50]

Unlike the other newsmagazines, the slightly more liberal *Newsweek* continued to keep the topic in full view up through the congressional hearings following the PSAC report. Coverage began, "It was a session of 'Author Meets the Congressmen.'" The reporter likened a scene of senators requesting autographs in their copies of *Silent Spring* to "a Rachel Carson fan club." An accompanying cartoon originally run in the *New Yorker*, showing a suburban matron in a garden supply shop and captioned, "Now, don't sell me anything Rachel Carson wouldn't buy," was included in the *Newsweek* article because it had been introduced into the *Congressional Record*.[51]

Given the newsweeklies' status as journalistic publications devoted to impartial news reporting, readers might have been surprised to detect partiality in coverage of *Silent Spring*. Some kind of slant, however, was generally considered more than acceptable in other national magazines' presentation of the debate, for they were positioned explicitly to offer evaluation or judgment on matters of public interest. Readers of *Reader's Digest, Saturday Evening Post, Life,* and *Redbook* (which in that era was a general-interest publication aimed at young adults) relied on editors to look into an issue like pesticide abuse, make a determination about it from a (sometimes vague) level of expertise, and then write about it in an accessible, "digested" way. The editors felt called on to present these issues in a personalized way, even while sounding a warning. The point was less to announce a debate or even to replicate it for the readers' own evaluation than it was to extract the editors' own ideas from the debate and present them for the enlightenment of the readers.

At the time *Reader's Digest* had by far the largest circulation of any magazine in America, at approximately 13.6 million in 1962.[52] As literally a digest of writings on the important or interesting issues of the month, the magazine presented a carefully edited collection of articles ostensibly drawn from other publications of the day. Editorial commentary was never included, but conservative politics and philosophy were occasionally discernible in the selection and editing of pieces. Some charged that the *Digest* covertly developed its own stories to be planted in selected magazines for later reprinting in the *Digest*; Harold Ross of the *New Yorker* concurred with those charges and found this practice particularly objectionable. On those grounds, Ross declined to work with the *Digest*, which meant from the start that the *New Yorker* series would not appear in the *Digest*.[53]

The *Digest* had previously run articles both touting and questioning the benefits of insecticides; but on the whole the emphasis had been on wonders of modern science in combating insects. Although *Digest* editors had declined Carson's proposal article on the dangers of DDT in 1945, in 1958 she contacted the magazine again on hearing of the magazine's plans to run an article on the benefits of aerial spraying. Carson wrote a long, detailed letter to owner DeWitt Wallace, offering her expert assessment of the "tremendous dangers" of spraying and urging him to reconsider. A senior editor responded, saying only "We shall weigh all facts."[54] Her plea appeared to have been heard, however, for in 1959 the magazine published Robert Strother's report of alarming examples of the harm of aerial spraying.[55] Two years later, however, in May of 1961, a "Reader's Digest Report to Consumers" featured Frank Taylor's "Good-By to Garden Pests?" which touted the great benefits of DDT, lindane, malathion, chlordane, and the like, naming the chemical companies responsible for their development (DuPont, American Cyanamid, Velsicol, Esso, Shell, Dow, and Union Carbide) and emphasizing the great contribution of the chemical industry toward realizing "the lazy homeowner's ultimate dream of a garden as a bit of paradise that will take care of itself."[56]

Perhaps not surprisingly, the *Digest* chose to reprint *Time* magazine's highly negative article as its own offering on the *Silent Spring* debate. An unnamed *Digest* editor appalled by the choice let Houghton Mifflin's Paul Brooks know that "the decision [came] from the top."[57] The *Time* article was run as if it had been a formal review rather than a "Biology" section report, now with the review-style notation of publisher and price inserted after the title. Although the editorial cuts in the article favored neither Carson's critics nor Carson, the subhead below the main headline read: "A new and hotly controversial book, *Silent Spring*, has alarmed readers with its grim predictions of death and desolation. Here are the reasons why many scientists strongly disagree with the author."[58]

In addition to probable sympathies with the corporate groups involved (listed in Taylor's 1961 article) and antagonism stemming from the *New Yorker's* traditional refusal to work with the *Digest,* some unpleasant negotiations occurred between Houghton Mifflin, Marie Rodell, and the *Digest* regarding a condensation of *Silent Spring.* The amount the *Digest* offered was deemed far too small both symbolically and financially, and the *Digest* refused to increase it. More than a year later, in October 1963, when the *Digest* later provided its own review of the debate, it warned that the "emotion of the moment" should not result in sacrificing "proven benefits because of unproven fears" or handicapping research with "red tape and excessive regulation." And in an intriguing swipe at the eastern establishment—presumably, the *New Yorker* and Houghton Mifflin in particular—the authors suggested that the "tumult" concerning chemical pesticides had been created by "angry people in the eastern United States."[59]

Given the corporate relatedness between *Time* and *Life*, one might predict that *Life's* attitude toward *Silent Spring* would have been as negative as *Time's*. In addition, Carson had already had uncomfortable dealings with *Life* over a 1956 article that was ultimately abandoned. On the other hand, in May of 1962 *Life* had presented a "Color Spectacle" photo-essay on the quest for alternatives to chemical pest controls just weeks before the *New Yorker* series appeared.[60] Moreover, *Life's* charter concern with

Photo from *Life* magazine feature, October 12, 1962. Photograph by Alfred Eisenstaedt/ Time & Life Pictures/Getty Images.

personality inevitably prompted editorial interest in the person of Rachel Carson. Pictorially, the article "Gentle Storm Center" was kind to Carson, showing her seated with her microscope, on a "lunch-hour bird walk" with a group of Audubon Society members, playing with her cat, and talking with children in the woods near her home. But in fact, the overall effect of the visual portrait is almost so benign as to be condescending and to undermine the authority of her research by associating her mainly with birders and those who doted on animals and children. Following the pictorial essay was *Life*'s own review of the issue, which outlined and evaluated her arguments and then presented manufacturers' rebuttals to what they deemed her "overstated" case. Overall, the text had more in common with the disapproving *Time* article than with the fond tone of its own accompanying photo-essay[61]—a fence-straddling subtlety uniquely available to the pictorial magazine format.

Bad feelings of a much more personal nature were quite likely behind the exceptionally antagonistic "Myth of the 'Pesticide Menace': Concerning *Silent Spring*" written for the *Saturday Evening Post* by Edwin Diamond. Diamond was the *Newsweek* science editor originally hired to work in collaboration with Carson on research for the book; when their perceptions of his role diverged, Diamond may have resented either being treated as a research assistant or the relatively peremptory termination of their arrangement. His sour challenge to her was at least clearly identified as opinion by its location in the *Post*'s "Speaking Out" section. The caption under his picture indicated him as having "worked with" Carson but implied that a fundamental difference in philosophy rather than roles had driven the split.[62] His extended comments on *Silent Spring* were perhaps the most contemptuous and stridently derogatory of any in the mainstream press, and he attributed the media attention to public gullibility and paranoia, which he compared to McCarthyism. Why the *Post* should have elected to make Diamond's diatribe its sole contribution to the discussion is unclear. But the fact that the *Post* ran Diamond's piece a full year after the book's publication underscores that the magazine saw its function not as reporting news but rather as elaborating on and ultimately participating in the public discussion before its designated segment of the public.

Although the large-circulation general-interest magazines may have indulged in varying degrees of subjectivity, taking a position on the book or the issue was definitively required of small-circulation specialty magazines. The specificity of these magazines' discussion of *Silent Spring* tended to take a rather personalized form, further emphasizing an immediate and individualized locality in the conduct of the debate. Moreover, the discussion in such specialty magazines allowed their readers to experience the debate within a manageably small node of the public sphere while sustaining awareness of the broader and more complex discussions in the general public forum.

If a magazine had ever investigated or taken a stand on pesticides, its entry into the *Silent Spring* debate as a participant-advocate was guaranteed; but *Silent Spring*'s arguments changed few editorial minds in either direction. This was true of the popular-science magazine *Scientific American*, the American Medical Association's *Today's Health*, the National Audubon Society's *Audubon Magazine*, and *Science*, official publication of the American Association for the Advancement of Science (AAAS).

Science had published articles on pesticides, DDT in particular, at regular intervals in the postwar era, the majority of which were concerned with specifics of the science involved or the efficacy in killing pests. Given that its parent organization, the AAAS, represented the scientific community, one could predict some resentment over Carson's implications that the scientific community had been lax or co-opted by industry priorities. Writing for *Science*, I. L. Baldwin crafted a review that became a standard enclosure in opposition mailings. Baldwin scolded, "I cannot condone . . . the sarcastic and unjustified attack on the ethics and integrity of many scientific workers."[63]

Other science-oriented magazines faulted *Silent Spring* for other reasons. William Vogt had expressed regret to Brooks that his review for the American Museum of Natural History's *Natural History* magazine could not be entirely positive, in part because he had been contacted by the World Health Organization in alarm that *Silent Spring* might undermine anti-malaria campaigns.[64] Although Vogt generally applauded Carson's mission and was not bothered by its advocacy ("She has been accused of being 'one-sided,' as though this were a fault. I have never heard St. Paul criticized for not giving Satan his due"), his measured review found her "vulnerable to attack because of a tendency to exaggerate and an occasional, uncritical acceptance of data."[65] *Scientific American*'s more positive review still faulted Carson's approach to science, accusing her of failing to acknowledge ecology (then a relatively new field) as the proper field in which to find answers. Not surprisingly, the reviewer, LaMont Cole of Cornell, was himself an ecologist and spent most of the review arguing his own take on some ecological specifics, such as insect resistance to chemicals (which Carson had, in fact, discussed). Otherwise he applauded *Silent Spring* despite its partisanship and made particular note of the fifty-five pages of reference "so the reader can look up the original accounts and judge the evidence."[66]

Magazines that were geared to business and farming interests focused on chemical industry responses and tended to draw heavily on material from industry trade journals. Zeroing in on business's core concern—the threat of government intervention—*Business Week*'s September 1962 article "Are We Poisoning Ourselves?"[67] recalled actions following the 1959 cranberry scare; and its two subsequent articles in May 1963 focused on the PSAC report.[68] *Farm Journal*'s "You're Accused of Poisoning Food"[69] often accompanied Baldwin's *Science* review in opposition publicity packets. But magazine-reading farmers did not necessarily share the corporate perspective, given the personal hazards they faced in their own handling of pesticides. The widely read monthly *Farm Magazine*'s book review, which did not appear until spring of 1963, offered a lengthy synopsis of the book with essentially no mention of the debate. Rather, it weighed both strengths and weaknesses and echoed Carson's wish for more study directed at better accommodation of nature.[70]

Audubon Magazine might have been expected to celebrate the victories of a former contributor in a long-cherished cause, but its coverage was rather muted. Other than reprinting two chapters in its September and November issues respectively, its discussion of *Silent Spring* tended to place it in the context of the society's ongoing effort to raise consciousness and change policy about pesticidal hazards. The society may have been down-playing its connection to Houghton Mifflin: Brooks was an active member and former officer, and Audubon Society president Carl Buchheister had been helpful in the preparation and promotion of the book. The organization may also have considered distribution of promotional and educational material the better way to draw members and the general public into the discussion.

Other nature and conservation magazines, including *National Parks Magazine*, were inclined to offer tacit agreement with Carson by presenting excerpts from the book to let her words speak for themselves. *True* magazine readers were also treated to a much-condensed version of *Silent Spring* in its April 1963 issue, bannered on the cover and introduced with Justice Douglas's phrase, "the most important chronicle of this century." *Field & Stream* had published two articles about pesticidal harm to wildlife in 1959 and 1960; and in September of 1962 the magazine presented "The Plight of the Woodcock," recording the ill effects of chemical pesticides dieldrin and heptachlor.[71] Though neither Carson nor her book was named in the article, its arguments were made using many points much as she had.

By contrast, a feature in the November 1963 *Sports Illustrated*, a Time, Inc., publication, shared *Time*'s derogatory opinion. "The Life-Giving Spray" lauded the role of pesticides in making the world better for man and beast. Author/hunter Virginia Kraft presented anecdotal personal findings that thanks to pesticides, bug-free flora had nourished "the nation's healthiest wildlife crop in many decades," feeding even greater numbers of game animals. She averred that *Silent Spring* "made Nevil Shute's [post–nuclear holocaust novel] *On the Beach* seem almost euphoric by comparison." The accompanying unflattering photo of Carson with rain hat, binoculars, and simpering expression evoked the stereotypically alarmist "little-old-lady-in-tennis-shoes." But Kraft leveled equally sharp disapproval at Interior Secretary Stewart Udall for "secondhand scare stories," suggesting an anti-Udall agenda in general might have been at work.[72]

Periodicals targeting home and family generally treated the *Silent Spring* debate as an opportunity to educate their readership—now postwar consumers—about pesticides. *Consumer Reports* had already advertised a special two-dollar paperback edition of *Silent Spring* offered to subscribers only; but the first mention of the public debate within the magazine itself was a January 1963 review detailing industry's countering media tactics.[73] Some readers wrote criticizing the partisanship and the "alarmism" in the *Consumer Reports* coverage. The magazine soon ran a responding article titled "The Public Needs to Be Alarmed," but it also offered a two-part article instructing readers on pesticide safety.[74]

Garden magazines, long supported by chemical company advertising, traditionally carried many articles on choice and use of pesticides. Their tendency now was to dismiss the basic questions about environmental dangers and to focus rather on education about proper pesticide use. Gardeners and home owners who considered insects and weeds their archenemies now read in the November 1962 *Better Homes and Gardens* that they should not be "panicked into throwing out all [their] dusts and sprays" *unless* they were "going to use them carelessly."[75] *Good Housekeeping* delayed comment until June 1963, finally telling its readers (with no mention of the book) that although pesticides are poisons, they posed little danger under existing guidelines.[76] A follow-up article in August reported (erroneously) that the PSAC report had validated what *Good Housekeeping* had said in June; but the editors added comment that reports of harm indicated carelessness by home owners and that therefore they wished to reemphasize warnings to be especially careful.[77]

Among publications concerning personal health, some similar finger-shaking at consumers took place in the American Medical Association's popular-medicine periodical, *Today's Health*, for which Frederick Stare occasionally wrote articles decrying food fads. His September 1962 blurb on "Spotting the Food Quack" implied that those claiming "our food is . . . filled with poison" were motivated by the desire to sell something.[78] In February 1963, another writer's "Pesticides: Facts, not Fears" featured garish illustrations of "enemy" insects and emphasized the benefits of pesticides, quoting opposition spokesmen such as Decker and Baldwin, along with the *Time* article, calling Carson's views nonsense tainted by atavistic worry about the balance of nature.[79] The AMA had also been the source of a number of anecdotal statistics used by her critics, including those cited during the congressional hearings comparing deaths related to pesticides, aspirin, and "even" salt. Stare's influential activities in the AMA's nutrition division and the fact that at the time the association had a politically conservative leadership may account for the absence of any AMA health warnings about pesticide use, even after the controversy had died down a bit. *Prevention* magazine, in contrast, published a highly laudatory review of "Rachel Carson's Masterpiece" in October, in which the *New Yorker* articles were counterposed against what the reviewer called *Reader's Digest's* "unfactual attack on organic gardening."[80] A better demonstration of cross-media consciousness and interaction could scarcely be offered.

Literary and intellectual magazines, despite their lofty detachment from popular, mainstream media skirmishes, were nonetheless susceptible to the same cross-media influences, as well as to personalizing their own contributions to the debate. Their purview was the domain of ideas more than advice on pesticide use, and the hallmark of their reviews and articles on *Silent Spring* was the importance of the authors' own thoughts and experience. For these writers, *Silent Spring* functioned mainly as point of departure for their personal exposition; and for them the goal of writing was as much to exercise their own authorial abilities as it was to appraise Carson's. The intent of their writing was to locate *Silent Spring* within the spectrum of all thought and literature, interpreting its intellectual significance and assigning its cultural importance. Often the rise of a debate provoked by the book was offered as evidence of status, and the worth of the debate was itself debated.

James Rorty of the *Commonweal* was so impressed with the book and so invested in his response that he sent a draft copy directly to Carson, bypassing the mediated distance between them.[81] Even more than *Scientific American's* Cole, he stressed the field of ecology as the proper locus for study and restated Carson's points in his own flowing prose. It was not enough to embark on new research: "we must choose life, the fragile, tremulous web of life that binds man to the earth, and reject the sprayed death that shatters this web. Let us pray that we make that choice soon."[82] Marston Bates was likely chosen for the *Nation's* review as author of such books as *Man in Nature* and *The Prevalence of People*, and he similarly placed *Silent Spring* within the theme of human arrogance in the face of nature. He seconded Carson's challenge to the immorality underlying pesticide policy and practices: "No one is in a better position than Miss Carson to arouse the indignation of the public and the conscience of the chemical industry, and it may well be that she has made a real contribution to our salvation."[83] At the *Atlantic*, Edward Weeks was more inclined to bear personal witness to the follies of pesticide use, recounting some sad incidents such as the silencing of neighborhood birds and the death of "every trout, eel, and frog" in a favorite fishing spot. Actual discussion of the book was reserved for the latter half of his review, and it too concentrated on specific bad consequences, drawn from Carson's record.[84]

The *Atlantic* was owned by Houghton Mifflin, which may well have played a part in its support of *Silent Spring*, although the magazine's East Coast, literary orientation was equally relevant. Whichever the case, nine months after Weeks's review in the magazine, a much longer and more thoughtful article by conservationist Clark Van Fleet appeared in the July 1963 issue, expressing a felt need to rally to Carson's cause. Recounting the attacks by Carson's critics, especially Nutrition Foundation head C. G. King, White-Stevens, and other members of the chemical industry, he worried: "Miss Carson's image has been seriously tarnished by this barrage of criticism. My own feeling is that she needs some unqualified support." His personal contribution had been to "do a bit of research on my own," particularly concerning the situation in the Far West, and he found that things were dire. Calling for limits and controls, he expounded in Carson-esque vein that "if the merchants of death have their way, we will soon have no birds at all to sing over the graves of our hero dead, no birds to fill the spring woods with their gaiety and chatter, and our land will be desolate and silent as a shroud."[85]

Another literary magazine by charter, *Saturday Review*, had an established orientation with respect to nature. Its editor, John Lear, was a nature writer and nature lover. Immediately after the appearance of the *New Yorker* articles, Lear had made a point of requesting a review copy of the forthcoming book.[86] When the Supreme Court denied appeal in the Long Island lawsuit, the magazine reprinted Justice Douglas's "Dissent in Favor of Man"[87] in its entirety. Later, on the release of the PSAC report, Lear himself wrote a "Personality Profile" of Carson in which he praised the report as a counterattack by the Kennedy administration on Carson's opponents.[88] In *Saturday Review*'s formal review of the book, a well-known author in science history and friend of Carson, Loren Eiseley, began as so many others did, with personal anecdotes. His further contribution was a history of the development of pesticides and the varying philosophies behind their use (including a parallel to thalidomide), within which he situated Carson and recapitulated her arguments. His concluding admonition read: "*Silent Spring* should be read by every American who does not want it to be the epitaph of a world not very far beyond us in time."[89]

Most tuned into the debate itself was the *New Republic*, whose short August editorial "Buzz, Buzz, Buzz" was among the earliest acknowledgments that a media debate was under way. It noted the responses already published in *Chemical Week* and *Chemical and Engineering News* and remarked on a peculiar characterization in *Agricultural Chemicals* which called the earlier *Life* article on biological pest control "the start of the 1962 campaign to stir up the public against pesticides."[90] In its August 1963 report on the congressional hearings, the *New Republic* said that the opposition had been sending out newspaper feature stories for many months stressing the positive side of pesticides while "the initial impact of Rachel Carson's *New Yorker* articles (later published as *Silent Spring*) has worn off" and that therefore "Congress is unlikely to act."[91]

The observation that coverage in such intellectual and literary magazines customarily took the form of "mere" book reviews is thus deceptive insofar as the world of literary reviewing implies lofty detachment from the mainstream. As personalized as the reviews were—and as consumed with the reviewer's own thoughts, experience, and rhetoric—what was often being reviewed was the debate itself, at least as much as the book. The "reviewers" were often well aware of reportage and commentary in other media and frequently were very concerned with what the real-world upshot of the debate might be.

Broadcast Media and CBS Reports

No media, however, were more cross-media-conscious than those of broadcasting, yet no media left so little record of their own role in the *Silent Spring* debate. That radio and television were involved in the *Silent Spring* debate can be known largely through notices in print media, since television and radio have not archived their own history well, particularly concerning their news and editorial activities. Other than passing mentions of local broadcasts in newspaper articles, broadcast media's activities must be deduced from archived correspondence and records.

Regarding radio, for instance, the sense that local stations in the Northeast and Midwest were the most actively involved in discussions of *Silent Spring* can be drawn from memos, from comments in preserved newspaper coverage, and particularly from readers' letters. For example, one supporter wrote to Carson that he had discussed *Silent Spring* "and its implications for one hour on the local radio station here in New Haven last week and had an excellent response on a program called, 'Face the People.' Listeners telephoned in questions which I immediately answered on the air."[92] Many radio stations already had book review programs, and many organized local debates within farm, garden, health, or public affairs programming.

Conduct of the debate over the airwaves can also be inferred from such records as the frequent requests for Carson's participation and White-Stevens's scheduling of numerous appointments with local radio stations during the summer and fall of 1962. Nationally, whatever went out via the Associated Press wire service was also available to its radio service. Additional crossovers involved the rising number of multimedia personalities, such as Erwin Canham, who was both *Christian Science Monitor* editor and moderator of a national television public affairs program, and especially Eric Sevareid. Sevareid discussed *Silent Spring* twice in his syndicated newspaper column, was host of the *CBS Reports* television program, and made his weekly radio column a podium in the debate as well.

By contrast, relatively little record exists of coverage on local television, although CBS's contractual restriction of other television appearances by Carson should be remembered here.[93] Nonetheless, that need not have prevented local stations from producing their own shows on the topic assuming they had the resources; and the scarcity of references to other television broadcasts, nationally or locally, does not necessarily mean that none occurred. In fact, CBS had purchased some film footage of poisoned wildlife from Jack Woolner of the Massachusetts Department of Fish and Game, who had recently made a film on the ravages of DDT for local television station WHDH.[94] Generally, however, local resources were not great, even though national television was experiencing a "golden age" of documentary reporting. It was therefore the national networks that turned television attention to *Silent Spring*, although not until much of the discussion had already taken place in the print media.

True, ABC News did rebroadcast an earlier program on the trials of a crop duster's life, including illness from pesticides, in August 1962 as the debate was gathering momentum.[95] And in early November 1962, the young public television network, National Educational Television, televised the Canadian Film Board's 1960 two-part program "Poisons, Pests, and People."[96] The press release sent to American television-page editors explicitly linked it with *Silent Spring* and promised a "sober and thoughtful look at the entire problem."[97] The film was indeed two half-hour segments of sober, thoughtful, balanced, and fairly technical treatment, involving the Canadian counterparts of U.S. interests involved: a Cyanamid Canada spokesman; representatives of forestry, paper, and farming interests; scientists; doctors; officers from government departments; and a representative of the Canadian Audubon Society. The program even zeroed in on the 1959 U.S. cranberry scare with rare footage of HEW Secretary Arthur

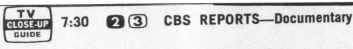

TV CLOSE-UP GUIDE 7:30 ❷ ③ **CBS REPORTS—Documentary**

"The Silent Spring of Rachel Carson." Every year, insects inflict damages on U.S. property, crops and forests amounting to an estimated 14 billion dollars. Man retaliates with chemical pesticides. However, scientist-author Rachel Carson, in her recent book "Silent Spring," charges that these poisons may endanger people as well as pests, food as well as the insects that attack it. Tonight's film charts the extent of chemical war on pests, outlines present controls on pesticides and offers opposing views in the bitter controversy "Silent Spring" has engendered.

Miss Carson, in an interview with writer-producer Jay McMullen, reads portions of her book. Cameras cut away to show the effects of pesticides on fish, birds and insects.

Secretary of Agriculture Orville Freeman and Dr. John Buckley, U.S. fish and wildlife research director, are among Government spokesmen heard. For in-

dustry: Dr. Robert White-Stevens of American Cyanamid's research division. Eric Sevareid is the reporter. (60 min.)

'SILENT SPRING'

TV GUIDE   A-59

TV Guide listing, March 30, 1963. Reprinted with permission from TV Guide Magazine Group, Inc., © 2005 TV Guide Magazine Group, Inc. TV GUIDE is a registered trademark of TV Guide Magazine Group, Inc. Yale Collection of American Literature, Beinecke Rare Book and Manuscript Library.

Fleming announcing his recommendation not to buy northwestern cranberries. As balanced as the program was, it could have functioned in the debate similarly to the much-delayed CBS program had it been an American production and had its style been considerably more lively than the dry, "talking-head" format it used. But it lacked the engaging presence of Carson for her side of the argument, and it was broadcast still early in the energetic media drama that would climax with the landmark *CBS Reports* spring 1963 broadcast, "The Silent Spring of Rachel Carson."[98]

Both CBS and Houghton Mifflin had seen to it that television editors and columnists were well aware of the forthcoming program. *TV Guide* featured a "close-up" in the program notes,[99] and ads were placed in the major newspapers. When some advertisers withdrew their commercials, news of their action added to the anticipation; and CBS's ads for the program[100] appeared in the *New York Times* on the same page as the *Times* news report of the sponsors' withdrawal.

On April 3, 1963, the *CBS Reports* program opened with juxtaposed statements from Carson and White-Stevens. Carson's voice was heard over footage of spraying aircraft and dying animals and birds: "Can anyone

Newspaper advertisements for *CBS Reports*. Yale Collection of American Literature, Beinecke Rare Book and Manuscript Library.

believe it is possible to lay down such a barrage of poisons on the surface of the earth without making it unfit for all life? They should not be called 'insecticides' but 'biocides.' " Speaking directly into the camera, a white-coated, bespectacled White-Stevens countered: "The real threat to the survival of man is not chemical but biological. . . . If man were to faithfully follow the teachings of Miss Carson, we would return to the Dark Ages, and the insects and diseases and vermin would once again inherit the earth."[101]

After one of the few commercials to interrupt the broadcast, Sevareid returned to *Silent Spring*, comparing it to *Uncle Tom's Cabin*, and then proceeded to present in turn the perspectives of industry spokesman White-Stevens, USDA Secretary Orville Freeman, a citrus grower, the FDA's George Larrick, two representatives from the Public Health Service, one from Fish and Wildlife, and a representative from the PSAC. In carefully calibrated balance, substantial time was devoted to Carson herself, interviewed directly and also as "voice-over" reading passages from her book against scenes of nature in beauty and in distress. Approximately twenty minutes into the hour, Sevareid turned attention directly to the impact of the debate in the public eye, quoting Lee's *New York Times* report of the opposition's earliest actions and the suggestion that Carson might deserve the Nobel prize, along with *Time*'s negative comments. Then, and at intervals throughout the program, collages and close-ups of headlines, pictures, and articles from a myriad of newspapers and magazines defined the recurring visual themes of the media controversy.

As the broadcast neared its end, CBS made an assertion of studied impartiality and research-based authority, seeking to set itself somehow above the entirety of the cross-media debate. Sevareid turned the commentary over to producer Jay McMullen, who emphasized the long period of research that had gone into his investigation. McMullen added, "Eight months ago, we set out to determine just how serious the pesticide problem really is. In that attempt we have failed [due to] the appalling scarcity of facts." Although the true reasons for the delay from the original November broadcast date are unclear,[102] the emphasis on his research efforts can be seen as the television correlative of the book's fifty-five pages of documentation, lending weight and authority, even if McMullen found the quantity of facts insufficient to reach an objective conclusion. Then, in the same way that print reviewers sought to impose their individual and personal flavor on the debate, the program concluded in typical Sevareid style, with commentary on the philosophical issues at hand, primarily concerning "a conflict of attitude toward man's role in his environment, and his attempts to control and manipulate nature for his own benefit."[103]

CBS Reports was a newsmagazine airing every other Wednesday evening; thus it was stunningly fortuitous that the PSAC report was released on Wednesday, May 15, exactly six weeks after the April 3 broadcast. However, Sevareid and CBS almost certainly had a copy of the long report in advance of that evening's update, "The Verdict on the Silent Spring of Rachel Carson,"[104] just to manage production logistics. Even though a fair amount of the segment comprised portions of the first broadcast, Sevareid's lengthy summary of the report was followed by extended responses from Carson in Philadelphia and the National Agricultural Chemicals Association's Parke Brinkley—suggesting that they, too, had advance copies of the report. Carson claimed vindication and even observed that "there are several places too, in which the panel goes farther than I did." Brinkley's response was one of diplomatic disagreement, emphasizing the favorable aspects of the report and offering a promise of full cooperation in further study. Sevareid concluded by paralleling the intent and philosophy behind the book and the report, adding his own exhortation for more research and changes in policy and practice. CBS's broadcast pronouncing the PSAC report "the verdict" seemed to mark both climax and end point of the media debate, with an epilogue in Carson's death in April 1964.

Most obituaries cited Carson's lyrical nature writing and her crusading spirit, even as she was dying of cancer, and thanked her for her contributions to the growing interest in ecology. Not surprisingly, *Time*'s obituary (or rather its death notice in the "Ecology" section) asked whether her cancer accounted for unreasoning zeal in scapegoating pesticides; and it dismissed much of *Silent Spring*'s worth. Restating the charges of her critics "who included many eminent scientists," the notice further expressed relief at the defeat of certain government restrictions that were "so sweeping that if they had been passed and enforced, they might very well have caused serious harm." Nonetheless, even *Time* conceded that "Rachel Carson may be remembered for many a spring after the passions she aroused have subsided."[105] And indeed, media commentary and criticism have popped up from time to time since then, notably in 1972 with the government banning of DDT; in 1987, on the twenty-fifth anniversary of the publication of *Silent Spring*; and even in the present day, in the context of bioengineered crops and insect-borne disease epidemics.

Conceivably, the PSAC report would have been treated as end point and resolution by other media even without the CBS broadcast. Perhaps after almost a year of controversy, Carson's critics had recognized that their battle against regulation was now best waged out of public view; and perhaps the attention of the uncommitted public had turned elsewhere. In the interest of providing full perspective, in fact, it may be important to put that final CBS broadcast into the larger historical context. On the day the PSAC report was released, astronaut Gordon Cooper was attempting a dramatically dangerous reentry after earth orbit; and the *CBS Reports* for that evening had to work around live cut-ins. As the *Washington Post* responded to Houghton Mifflin publicist Anne Ford's query about headlines, "Astronaut pushed pesticide report to page 5. Sorry."[106] Yet the rhythm of the media coverage had reached an entirely natural climax in the two nationally televised programs, with the first defining the debate in immediately visible and audible terms, and with the second—broadcast soon enough that the audience's memory was not too taxed—presenting a reassuring, if governmental, resolution of the matter.

Book versus Magazine

CBS's reliance on Carson's book as both boiling stone and authoritative resource exemplifies better than almost any other instance the synergistic relationship between books and the other media. But CBS had also relied on the discussion in the print media at large to illustrate the significance of the debate and justify the attention of the huge public eye that was national television; the audience for the *CBS Reports* program was estimated to exceed 10 million viewers. A magazine article, no matter how prestigious or how large in circulation, could not have commanded the same attention, and neither could an author of a single magazine article. Moreover, by itself, no such article would have had the intellectual and cultural weight to place an item on the media's agenda and keep it there for over a year.

The absence of Carson's disclaimer from the first *New Yorker* installment as well as the lack of bibliographic citations in the *New Yorker* series were at first as noteworthy to all the media as they had been to the opposition; and thus it also made considerable difference to the media that her arguments were soon to appear in book form. *New York Times* science writer Brooks Atkinson wrote to Ford: "Am I right that you are going to publish the Rachel Carson material in the autumn? The articles in the *New Yorker* have made a terrific impression, especially on me and my friends. I shall want to write about the whole subject in the autumn. I would appreciate anything you can tell me about her sources and how she gathered the material."[107] A few weeks before the book publication date, he wrote in the *Times* that "nothing in the field of conservation has provoked such an explosive response as Rachel Carson's articles [which] constitute part of a book of 297 pages. . . . As a certificate of scientific responsibility, the appendix consists of fifty-five pages of citations and references."[108]

The very fact that a book loomed but had yet to come out enlivened the media drama during the summer of 1962, especially regarding the mounting opposition. In a *Chicago Tribune* story headlined "A Book Not Out Yet Reaping Praise and Condemnation," Harry Hansen wrote, "The members [of the American Chemical Society] knew that Miss Carson was attacking the miasma of insecticides, and they were afraid she might influence public opinion adversely and put a crimp in the spraying business. Some of them had read part of Miss Carson's book in the *New Yorker*, and some had gained their prior knowledge by mysterious means."[109] Yet, again, it is clear that the dramatic opening provided by the first *New Yorker* installment and the returning volleys by Carson's critics provided early warning to the media of the possibilities for a full-scale and credible public debate on pesticides. The combined effect of the prelude in the magazine series and the book's complete statement established a continuing rhythm within which the media debate acquired strength and complexity as a public issue.

The Unread Book

As that "one formidable indictment," the book *Silent Spring* supplied the media not only with a sequence of debate-defining events but also with the twin star of author-and-book representing both argument and evidence—in effect, an essential, probative underwriting of the terms of the debate. In other words, the credibility and authority generally inherent in the book form and specifically conferred by Carson and her publishers on *Silent Spring* meant an important legitimation of the public discussion. Reciprocally, however, the media conferred an implied legitimacy on Carson's arguments simply by taking them, and their rebuttal, seriously enough for continued coverage.

This reciprocity underscores an important aspect of the relationship between books and the media, one so obvious as to be taken for granted and yet so overlooked that the two—books and the media—are rarely studied together: they share a mission of purposeful communication. If Carson sought to alert, inform, and mobilize the public by writing her book, then in reporting on *and participating in* the *Silent Spring* debate, the media had the same goal, in part or in full. Media discussion took the ready-made disagreement on a matter of public welfare and built it into a public debate waged at all levels, from community groups to the halls of government—always with the implied possibility of public mobilization to affect policy. If the mission of the press extended beyond simply informing to ensuring Habermas's public "rational-critical" discussions in a Miltonian "marketplace of ideas," debates like the one about *Silent Spring* provided the ideal situation for fulfillment of that mission. Yet if the *media*

We Have Not Read It

There have been many inquiries about our opinion of Rachel Carson's new book, Silent Spring. We have read several reviews but not the book, so have no firm opinion. It is being widely defended by wildlife conservationists and widely attacked by manufacturers and distributors of certain chemicals used in pesticides. This controversy is likely to continue and the book may prove to be the needed impetus for a full clarification of the subject. We read enough of it, as it appeared in The New Yorker, to fully realize its potential importance.

As one reviewer said: " . . . she does point out that man's ability to create new poisons has outstripped his ability to calculate the overall effect of their use." That extensive study is called for is clear. Meanwhile her book will force such study to be made.

Editorial, Lynchburg, Virginia, *News*. Reprinted with permission of The News, Lynchburg, Virginia. Yale Collection of American Literature, Beinecke Rare Book and Manuscript Library.

functioned to place the issue of pesticide use and abuse on the *public* agenda, it should never be forgotten that, after only passing previous notice, it was Carson's work that managed to place the issue so squarely on the *media*'s agenda.

To be sure, the media's own desire to catalyze, purvey, and participate in controversy took the debate much further than could ever have been the case had the discussion been limited to the book's purchasers and readers. Although the single, identifiable book was always the ostensible core of the discussion, the media supplied a custom-tailored context for each pocket of audience and readership, a context that itself prompted elaboration and embellishment. As the media carried the debate into the enormously complex levels and segments of the audience—local to national, general to specialized—something of each level and each segment was added to the terms of the debate; and in that sense the ramifying power of the media was far more than simple amplification of the original voices of Carson and her critics. Second, as the media performed that amplification and ramification, the arguments of each side of the debate were potentiated, gaining energy, strength, and significance, in turn having the same effect on the debate itself.

Finally, in the process of fostering, carrying, and joining the debate, the media could provide their audiences with a lively overview of the issues, subject to constant revision and elaboration. For the audience, thereby, the media-borne debate offered a digest of the issues that made it entirely possible to have a reasonably good understanding of the issues *without* reading the book, or for that matter any of the opposition materials. Indeed, the implied probity that the book and the media conferred on each other gave the public tacit permission to discuss the book *regardless of whether they had read it or not.* What's more, some members of the media felt justified in commenting on the *Silent Spring* debate even while admitting that they themselves had not read the book. Among scholarly readers, the idea of discussing a controversial book without having read it carries the shameful taint of intellectual dishonesty. Within the culture of book reviewing, however, there is a traditional if rarely acknowledged assumption that those who read a review may never read the book and, furthermore, that debate about a book's topic may acceptably take place on the basis of the reviews and media coverage, rather than on a reading of the book itself. And at least one analyst of the book reviewing culture has made the candid observation that reviews can and do function as crib sheets for social conversation.[110]

As we move into discussion of the media audience's expectations and perspectives on *Silent Spring*, we must recognize that this book had a public life based on having been read and discussed in the media by a core group of writers and broadcasters, but we know not by whom else. The number of people who actually read their copies of the articles or the book is unknowable, and the number of people who knew about and had opinions about the controversy without ever seeing the text firsthand is even less knowable. Nevertheless, the existence of a significant group of letter writers and activists among the general public indicates not only that *Silent Spring* had a large public audience but also that many among that audience considered themselves an integral part of the debate.

CHAPTER 6

Audience: "This Ought to Be a Book"

According to legend promoted by Carson herself, *Silent Spring* was inspired by a reader's letter. In 1958 Olga Huckins had written to Carson describing the death of songbirds caught in a spray of DDT and asking Carson to look into it. Four years later, a note from a Houghton Mifflin copy editor moved by the manuscript was probably the book's first fan letter.[1] The same impulse to write and connect was felt by thousands thereafter who wrote (or sometimes called), not just to Carson but also to Houghton Mifflin, the *New Yorker*, CBS, newspapers, radio stations, government offices, and each other. For these letter writers, writing represented participation in the debate, whether privately or as citizens; and it also represented association with others in that debate. In actuality, Carson and Huckins had corresponded earlier, and Carson had been interested in the topic for more than a decade before Huckins's letter. To Carson, however, it was important that the public history of her book be closely linked to her readers' experience and response. Similarly, others saw letters and telephone calls as evidence by which Carson's publishers, her supporters, her opposition, and even the media could assess the success of their respective missions: Did the audience "get" the message? What did they think about it? Further, for the media in particular, the letters provided material for yet another round in the ongoing debate.

Although many thousands of people wrote directly and privately to Carson, what is of focal interest here is the great number who communicated to the *public* by way of communicating *in* the media or *to* the media themselves. A premise of my work has been that the *Silent Spring* debate was conducted in a species of public sphere located in the nexus of communicators, media, and audience; and I have used the more encompassing word "audience" in preference to the narrower term "readership." To close the theoretical circle of participants in the debate, we must try to consider the responses of the audience—including book, newspaper, and magazine readers; radio listeners; and television watchers—intending to join the *public* debate in one way or another. We should therefore cast our net not only over those whose letters actually appeared in print but—as best we can—also over those who expected, through communicating, to have an impact on both media and the public at large in the conduct of the debate.

Ideally, we should therefore include consideration of non-written communications, too; and a few reports of telephone calls and even personal interactions with the media can be found. One example is the call-in radio discussion reported by a Connecticut man who had been a panelist taking listeners' calls.[2] Such calls deserve to be considered part of the body of response, but records of them may not exist beyond sparse passing references. Moreover, the body of letters to, in, or about the media (including the *New Yorker* and Houghton Mifflin) is not only large in number but also a solid illustration of the audience's view of its own role. Fortunately, the letter is readily archived, albeit far less often than should be the case.

Even in a study exclusively concerned with public rather than private correspondence, however, access to letters both published and unpublished can be problematic, especially given the disinclination of the media to preserve their own history. Questions of privacy, proprietary material, and copyright complicate the matter further. There is also a serious risk in assuming that what audience-writers write is representative of the attitudes of the entire audience. The most anyone could claim is that what they write is representative of the views of those who cared enough to take up pen on the given issue, write a letter, and mail it. Finally, it is of central importance to acknowledge that looking at published letters to the editor may reveal more about the letters-page editors than it does about the audience at large. For that reason, we will begin by highlighting the role of letters-page editors as they probably affected the public's view of its own response to the *Silent Spring* debate.

The Role of the Letters-Page Editor

The common assertion of letters-page editors has historically been that an effort is made to create fairness and balance in the choice of letters to be published.[3] Yet it would be exceptionally naïve to suppose that manipulation does not occur. At the very least, a well-written letter will always have a greater chance of appearing in print than a garbled one. The number of letters printed may be more a function of space and news activity than of the number received on either side of a debate. And not all organs are necessarily committed to publishing letters in matched proportions to the number received pro and con on a given issue. Furthermore, despite the best intentions of the editor, the temptation to print an outlandishly foolish letter from someone espousing a cause disliked by the newspaper may be difficult to overcome. Even within efforts to balance representations of two sides, the choice of spokespersons can be conditioned by tangential or even ulterior motives.

Four instructive cases in point are the letters printed by the three Time-Life magazines, *Life, Time,* and *Sports Illustrated,* and by the *Saturday Evening Post* after their respective, negative articles on *Silent Spring. Life, Time,* and *Sports Illustrated* each observed a rigid format of four letters, two pro-Carson and two anti-Carson. But in all cases, the two anti-Carson letters used forceful language praising the magazine and condemning Carson, while at least one of the pro-Carson letters was actually faint or ambiguous in its support. Of the two anti-Carson letters in *Time,* one pronounced "a pox on authors who will pervert the truth for a few lousy bucks," and one of the pro-Carson letters contained the "admission" that Carson had used "Madison Avenue tactics."[4] *Sports Illustrated*'s antagonism toward Interior Secretary Stewart Udall was played out in the counterpoint between the second two pro-and-con letters, with the last word being a resounding repudiation of Carson and Udall together.[5]

The authorship of the letters was also a matter of careful arrangement. *Life* presented a scrupulously symmetrical double-pairing of two authorities—one pro and one con—and two "ordinary people"—also one on each side. The authorities were Parke Brinkley, head of the National Agricultural Chemical Association, and Roger Tory Peterson, renowned nature author and Audubon Society officer; the title and association of neither Brinkley nor Peterson appeared, although Brinkley's text began "Here at the NACA. . . ."[6] The "ordinary" writers were a farmer ("we farmers have no choice") and a woman who said only that there had been much indiscriminate spraying "to save that sacred cow, the American elm"—phrasing that played out a recurrent opposition theme about the (specious) choice between "trees versus birds." *Sports Illustrated* used letters from Roland Clement, whom it identified as "Audubon Society Staff Biologist," and a

University of Rochester doctor. *Time's* identified authorities were, as an exception, both in the pro-Carson camp: the president of the National Wildlife Federation and a Syracuse University researcher.

The *Saturday Evening Post* ran six letters responding to Edwin Diamond's acrimonious diatribe against the "myth of the 'pesticide menace.'"[7] The only identified authority was the author of the first letter in the column, none other than Carson's ubiquitous and staunch opponent Frederick Stare, whose degrees and affiliation with Harvard's School of Public Health were duly noted. Of the six letters, three clearly supported Diamond, two clearly supported Carson and condemned Diamond, and the last ambiguously compared paranoia and psychosis, leaving interpretation up to the reader—although Diamond's most adamant point had been that collective paranoia was the best explanation of *Silent Spring's* success.[8]

Of all magazines, the one that carried by far the greatest number of letters to the editor was, perhaps not surprisingly, the American Chemical Society weekly, *Chemical and Engineering News* (*C&EN*), which in the four months after the first *New Yorker* installment carried a total of sixteen letters. What may be surprising, however, is that of those letters, only three were critical of Carson. Three editorials in the same period had been devoted to the debate. The first was an admonition against name-calling, acknowledging that perhaps improvement was needed and calling for cooperation toward that improvement; the second had been a hand-wringing prediction that the prevailing emotional atmosphere would stymie research progress.[9] In the third, however, editor Richard Kenyon explicated the role of an editor in a controversy: "*Chemical and Engineering News* has expressed opinions on its editorial pages and has published reader opinions. Offering opinions in this fashion is done in the interest of stimulating thought on matters of importance to its readers."[10]

Although *C&EN's* coverage was hardly laudatory of *Silent Spring*, not all of it had been as resoundingly negative as William Darby's review in it, perhaps because of the mixture of academicians and practitioners in the society's membership. The majority of letter writers nonetheless took the journal and its writers—Darby above all—to task for being as biased and emotional as Carson was accused of being, or for playing as fast and loose with facts as they had accused Carson of doing. Some, however, took on other letter writers, as two university professors did in a joint three-column rebuttal letter that covered nearly all the usual criticisms leveled at the *New Yorker* articles (poor scholarship, unjustified conclusions, extremism in advocating the abandonment of pesticides) and included a three-paragraph parody extending the logic to the banning of antibiotics and chemothera-pies. In sum, the editor of this potentially partisan periodical was appar-ently compelled to produce a microcosm of the larger debate, reaching toward balance over time—something not seen among other magazines, which tended to confine the debate to one initial report and, at most, a single session of reaction.

Newspaper editors had the comparative luxury of allowing the letters-page debate to live or die according to news events. Logically enough, those that gave the issue of pesticides the most reporting and editorial space were also the most likely to print at least one letter on the topic. In the Midwest, however, notably in Chicago and Des Moines, some letters on pesticide abuse appeared before the newspaper itself took any official note of *Silent Spring*. In the *Des Moines Register*, three July 1962 letters protesting spraying projects[11] preceded any editorial comment or review of *Silent Spring* (other than the front-page editorial cartoon, reproduced in chapter 5), although the wire services and the *New York Times* were carry-ing notice of the reaction to Carson's articles, and the editorial staff was likely aware of them. Midwestern localities were already embroiled in squabbles about roadside spraying, Dutch elm spraying, and anti-mosquito spraying. Editors may have been using the letters to establish that what Carson had to say was "not new" as well as to connect what they expected to be looming controversy to their own local interests early in the debate's development.

Discerning what lay behind the editors' choice of letters for publication is, of course, a matter of hindsight and guesswork. Some letters directly contradicted a previous review or editorial, but in other instances letters and editorial stance were congenially congruent. The mere study of the physical layout of letters with respect to other letters, columns, editorial comment, and even advertising copy could suggest dynamics that could never be tested. Whatever the arrangement, the editor effectively still had the last word, so to speak—even when proxy for that last word was given to a letter-writing reader and even if that writer's word did not conform to the editor's own opinion. It was still the editor's choice to print the letter or not.

On occasion, though, that editorial last word could be had visibly, in an "Editor's Note" containing clarification or even rebuttal, as happened with a letter to the *Memphis (Tenn.) Commercial Appeal*. The letter had re-minded the book-page editor that any review of *Silent Spring* was one-sided unless it said that feeding the growing population without chemicals would be impossible. The Editor's Note replied that "before Rachel Carson's book appeared, several chemical companies had broadcast refutations," and it provided a full quote of Carson's reply, restating her declaration that "I don't believe we could stop using pesticides tomorrow." It then cited her suggestions for alternatives and set forth the fact that "we spend $1 billion a year on storing agricultural surplus."[12]

Finally, it is possible to think that the timing of the *Silent Spring* debate may have been as significant as other editorial considerations regarding audience response. The debate arose and gained quick vigor in the summer—traditionally a slow time for news. Such an inference is somewhat debatable in the case of the busy summer of 1962, which saw rising Soviet activity in Cuba, an accelerating space race, developing consciousness of the Vietnam conflict, and growing civil rights struggles. Moreover, that inference would be simplistic—especially in light of the observation, for example, that many of the published letters drew parallels to the tragic side effects of thalidomide and the dangers of nuclear testing fallout or even war with the Soviet Union. The specifics of timing, however, might well have figured in editors' decisions insofar as congressional and local elections loomed the following November.

Whatever the letters editors' orchestrations might have been, the editors could not have written the letters themselves (giving most the benefit of the doubt on that point, at least). They could not have requisitioned the number or content of letters from their public, nor could they have manipulated their purpose except in the most indirect way through previous or subsequent editorial and news copy. As we look at the nature of letters to editors about *Silent Spring*, both published and unpublished, we can see, therefore, some evidence of who was likely to be moved to write and what their expectations might have been about their letters' impact. In other words, despite the intervention or "gatekeeping" of the letters editors, we are still able to discern something about the audience's own perception of its role in the *Silent Spring* public debate. Their writing of those letters represented the action of engaged, responding recipients of Carson's and her opponents' messages, but more, it represented the action of those fully expecting inclusion as participants in the debate.

Audience Participation—Expectations and Assumptions

To begin with, public letter writing about *Silent Spring* can be placed in a special historical context with William J. Lederer's 1961 book, *A Nation of*

Sheep, an intriguing insight into an early sixties-era philosophy of activism vis-à-vis the media.[13] Author of an earlier best-seller on America's international image, *The Ugly American,* Lederer wrote this book ostensibly to answer the question "What can the average American do about the posture of the United States in foreign affairs?" especially "when the press is so convinced that the American people don't want the hard facts . . . that it makes only routine efforts to report them."[14] Although the bulk of his book was devoted to problems evolving in Southeast Asia and misinformation by government and press, the third section provided "specifics" of what to do "at a national level" and what "at a personal level" to counteract ignorance and consequent apathy. He counseled self-education and skepticism toward the media, and he specified particular actions to be taken. His blueprint for mobilization had at its core the need to exert pressure on the media and government, using individual and community-based letter-writing campaigns to demand more straightforward information and better coverage of the information and the issues.

A Western Electric employee writing a letter of praise to the *New Yorker* for the *Silent Spring* series enclosed a condensation of Lederer's book, which had been circulated internally by the company; and he quoted its final statement:

> Whenever you object to something; whenever you have a constructive way to improve matters; whenever you want to know the facts of an issue—write, call or telegraph. Write the President, your Congressman, Senator, newspaper, your radio and television station. . . . A citizen must perform a citizen's duties every day. . . . Now—as you read the last page—you can reach for your writing materials, for your newspaper, for the telephone—and take the first step of personal action.[15]

Although this idea of civic action was not new (neither is there any special reason to think that very many other readers of *Silent Spring* had read Lederer's book), the underlying activist theme was echoed in the words and actions of the letters written about *Silent Spring* to a startling degree. Further, this similarity of objective and theme—that is, a familiar sense of mission—ran throughout responses to *Silent Spring*, regardless of letter writer, addressee, or expectation of publication.

Sources and Targets

The first questions to ask about these responses concern from whom they came and for whom they were intended. Arriving at a formal demographic profile of all 1962–63 respondents is, inevitably, problematic. In 1972 *New*

York Times letters editor Kalman Seigel cited a somewhat earlier study estimating that 46 percent of 10,000 letters to the editor could be identified as coming from men—most of them professionals, officials, or politicians of some sort, while 37 percent could be identified as coming from women, of whom 42 percent were "spokesmen for various movements and groups" and the rest teachers, "working women," or "housewives."[16] Regarding the general profile of those whose letters were published by the *Times*, Seigel cited a 1965 *Journalism Quarterly* study finding that published letter writers were "mostly elderly and middle-aged, predominantly well-educated; male by a three-to-one ratio; . . . residents of the area in which the publishing newspapers circulated; extremely well-read, and remarkably conscious of their voting privilege."[17] Although Seigel estimated that about 7 percent of all letters received by the *Times* were actually published, he did not say whether the demography of letters received was reflected exactly in the letters published. To assert that the same kinds of demographics applied to letters about *Silent Spring* would be foolhardy; but in a general way, the description has some relevance, particularly given that the *New York Times* printed more letters about the book than any publication other than *Chemical and Engineering News*.

The same profile may not have been strictly applicable to the people writing to other publications; but because of the manageable size and discrete definition of the collection of letters to the *New Yorker*, it is possible to examine the demographics of at least that letter-writing population.[18] Even though as a matter of policy at the time the *New Yorker* did not publish letters to the editor, letters were still written to the editor; and they constituted a response to the medium (rather than to the author). Of the letters written to the magazine about *Silent Spring*, the writers were slightly more often male than female. Almost a quarter of the correspondents wrote on letterhead stationery, specifying their titles or identifying themselves as academics, government or corporate employees, or officers of garden or conservation clubs, particularly local Audubon societies. More than a quarter wrote from the Northeast, but almost as many came from the Midwest and Far West, respectively, with the remaining quarter from the mid-Atlantic and southern states, Canada, Europe, and elsewhere. It is unlikely that this distribution exactly mirrored the geographic distribution of the *New Yorker* circulation;[19] and the disproportionately large representation from the Midwest and Far West is plausibly explained by the letter writers' frequent mention of agriculture and conservation efforts in their areas—as well as by the fact that several of pesticide-related catastrophes described in *Silent Spring* had taken place in the Midwest and Far West.

Otherwise, certain impressions emerge from a review of published and unpublished letters sent to newspapers and other magazines. First, the gender ratio and frequency of inclusion of titles were similar to those in the *New Yorker* group, but slightly more letters from men were published in national-circulation magazines and slightly more letters from women were published in local newspapers. More specific to the *Silent Spring* history was the question of the sympathies of those who wrote. The vast majority of unpublished letters to the *New Yorker* supported Carson. The fact that 98 percent of the letters written to the *New Yorker* praised her or the magazine was indeed remarkable enough to merit media notice, thanks in part to efforts by both Carson and her agent, Marie Rodell, to insert that very statistic into media discussion. Calculating the pro-con proportions in the print media is less fruitful except on a publication-by-publication basis, first because of variations in editorial orientation toward Carson, pesticides, industry, and the environment. Second, estimation is further complicated by the permutations and combinations of letter-publishing practices—for example, whether multiple letters on the same subject were likely to be run, and if so, whether they would appear together or on different days or in different sections.

The professional or social associations of pro-Carson writers whose letters were published were often identified—as for example, with local or national garden clubs or Audubon societies. The associations or professions of published critical letter writers were less frequently identified. Whether the difference in identification was a function of editorial decision or the writer's choice is difficult to assess. Thomas Jukes of American Cyanamid wrote often but rarely on letterhead stationery except to contacts in the Department of Agriculture; others, like a senior congressional staff member,[20] took pains to say they wrote as private citizens and not in any official capacity.

The letter writers' choice of addressee was sometimes the actual intended target of their communication, but by no means was this always the case. The ones who wrote to the *New Yorker* specifically (and not to Carson in care of the magazine) were writing their responses to the medium, either as if it were the true originator of the message or in its capacity as conveyor of that message. Although a few of these wrote in the mistaken hope that their words might be published, most were not writing to other readers via the editor. They were writing to interact with the magazine—communicating their praise (or condemnation) or urging it to do something further. A few wrote with similar sentiments to Houghton Mifflin, although some members of the opposition (notably Jukes) wrote to question the publishers' motives or to challenge their authority. Those who wrote to the magazine or the book publisher were compelled by a desire to provide feedback on the performance of the medium itself. The recurrence of the term "public service" in so many such letters reflected their presumptions about that function for that medium as well as about the writers' own authority to act as members of the public whom the medium must serve.

When the addressee was the editor of a periodical, the intended recipient was almost always a composite of editor and readership, and perhaps others as well. (Even in the exceptional case of the *New Yorker*, many wrote with the obvious expectation that their letter would be printed.) In the process or guise of providing feedback to the newspaper or magazine about how well or poorly it was performing its "public service," the writer's actual purpose was usually to offer commentary meant for other eyes on the issue at hand. These letters could be written spontaneously, on the occasion of reading Carson's words or witnessing a pertinent local event. They could be written in response to an editorial, review, or article. Or they could be written in response to earlier published letters. Whatever the stimulus, the writer's desire was that his or her voice be heard—not only by the editor or reviewer to whom the letter was directed but, even more important, by other citizen readers—other gardeners, consumers, voters. The inherent expectation beyond *that* was that a citizen's attention and comments demanded the attention of those in power—politicians and practitioners—who were thereby in effect the *final* recipients, regardless of whether the writer was conscious of that intention and whether there was evidence that these final targets actually read the letters. The letter writer who chose to write to the editor rather than directly to the president did so out of the expectation that his or her voice would be magnified by association with fellow readers of the periodical. For such a writer, participation in the *Silent Spring* debate was an intentionally public act designed to have impact greater than his or her individual franchise as voter.

Feedback: Why They Wrote and What They Said

At the most basic level, all letters—whether or not intended for publication—were therefore bids to communicate at a larger-than-personal level, as well as conscious efforts to respond and/or inform in some corner of the public forum. The effect of the letter writers' communication was to complete, publicly, the connection from author to audience; but in the process it also established a sharing, collectivized connection with the rest of the audience via the media. The impulse to connect—with Carson herself, with publishers, editors, legislators, government officials, and, above all, with other members of the reading audience—flowed from two paradoxical objectives. On the one hand was an individualistic desire to stand out from the perceived mass of undifferentiated audience members; on the other was the wish to enjoy the validation of association with the larger group. Among those writing to dispute *Silent Spring*, these impulses were somewhat less operative, particularly when the writer was a chemical company official or scientist involved in pesticide-related research, writing officially or as a matter of self-defense. Yet even then, the practice of referring the reader to other authorities or to items in other media reflected a similar requirement of validation by association.

How these dynamics played out specifically in the *Silent Spring* debate is reflected in the various types of responses. Necessarily, however, distinctions among those types are once again artifacts of analysis: the letters themselves were endlessly varied with all combinations of motive, content, and effect. Yet even the simplest communication had an implied role in the overall machinery of the discussion. Many who wrote directly to Carson, the *New Yorker*, or Houghton Mifflin were merely requesting further information or asking for reprints. Although offers to *give* additional information are easily interpreted as gestures of reciprocity, the act of requesting information or reprints was itself a form of feedback—indicating that one's attention had been attracted and that enough attention had been paid to generate a need for replication or amplification.

A more direct bid to participate was common among letters submitted for publication, although it was also seen in letters to the *New Yorker*. In such letters, the writers offered their own applicable experiences, often recounting the catastrophic results of aerial spraying or misuse of pesticides—dying or dead birds and other animals, or even deaths or health problems in their families. As the most personal form of response, these letters were vastly more likely to be sent directly to Carson herself. Nonetheless, the "I myself have witnessed . . ." or "I am one of the victims"[21] theme ran through a number of published letters to the editor, commonly to introduce substantive arguments of the merits of one or the other side. A significant number were addressed to Carson's publishers as testament to the worth of having published her work, but many also carried an oblique bid to be published themselves. That offer was occasionally overt, as with a number of writers who submitted a piece of their own writing— offering anecdotes or philosophizing about humanity, chemicals, and nature—to the *New Yorker*, not because they expected an exception to the no-letters-published policy but because they expected the *New Yorker* to continue the discussion by using a reader's response as another article.

Often writers felt compelled to notify other readers (and editors) that a debate existed—that an organized counterattack was in progress or that the attention of other media was being drawn. A letter to a Connecticut paper reported, "During the past two months the pesticide industry has been preparing itself for the biggest attack it has ever had to face. . . . Newspaper editorials all over the country are praising Miss Carson's excellent work."[22] Regarding the opposition, Toledo columnist Lon Campbell reported a letter from a Mrs. Sinclair, who had pointed out to him that "there is a remarkable similarity in their arguments, their approach, and the cases they cite."[23] In this case the responding reader accomplished several goals at once, even as she served the columnist's own purposes in summarizing the debate for his readers: she got the attention of the media, which in turn relayed her observation, thus informing the readership about the tactics of the opposition, thus indirectly supporting Carson while contributing her own framing of the debate (albeit one nicely congruent with the columnist's).

Another common, related theme was gratitude to Carson and her book for not only sharing but also validating beliefs already held by the writer in relative isolation. One somewhat typical letter said, "The publication of Rachel Carson's book, *Silent Spring*, and the widespread interest it has aroused is a great satisfaction to those of us who have long been distressed by the indiscriminate use of chemical pesticides, fungicides, and fertilizers."[24] Such sentiments functioned similarly to those of reviewers and columnists who placed Carson's work in the context of their own intellectual history. Much more than that, however, these letter writers were trying to create an authoritative association with Carson and other like-minded souls, thereby fusing her authority with theirs, for mutual support. Letters describing a sense of having been liberated—thanks to Carson's crusade—from marginalization as "cultists," "faddists," "cranks," or "crackpots" convey vividly the sense of validation achieved through the book's publication as well as the compounding of a previously unheard "voice" against pesticide abuse.

From expressing gratitude for validation of views already held and shared, it was then a short step to acclaiming Carson as an advocate and even a surrogate against those whose actions or policies the letter writers deplored. Among the letters to the *New Yorker* and to Carson, a common comment was that "there is so little an individual can do."[25] For such writers, the author's voice stood for their own, not simply by dint of drawing together all like-minded souls but also because of the "bully pulpit" given the author of a newsmaking book. Although few would likely have thought about it in such terms, they were responding to seeing their own views well-argued in a book that would be read by many and that would command access to many other parts of the media system. The authority provided by Carson's stature as book author and the documentation of her research was, once again, a critical consideration for those celebrating her advocacy. "These things have long needed to be said by someone whose professional qualifications and reputation would protect him from accusations of faddism or hysteria, or simply from being ignored,"[26] wrote one woman to the *New Yorker*. But the most concise declaration of this function was expressed in a letter printed in *C&EN*: "When the sellers of pesticides for profit lay the burden of proof on the buyer, then I submit the mute common man needs an advocate with the power and persuasion of Rachel Carson to defend him."[27]

These letter writers were not in fact "mute." They were adding the force of their voices to hers while they thanked her for championing the cause. Even those to whom the issue of pesticide danger was news and who felt compelled to say in effect, "this is a terrible state of affairs and we should be grateful to Miss Carson" were acknowledging her advocacy and were, furthermore, assuming that acknowledging it publicly was intrinsically important. Thus in the minds of such letter writers, their connection linked author, reader, and media reciprocally; and it drew both media and other members of the general public into a complex political process more dynamic than the simple model of two sides engaged in a public disagreement.

By far, however, the single most common impulse among the letter writers was to evaluate and to take sides in an arena both implicitly and explicitly public, with strong overtones of civic action. Thus responses often included judgment of the media's role part and parcel with Carson's. Even before the debate became defined as such, letters to the *New Yorker* effectively took sides, with the vast majority writing to praise Carson, the magazine, or both, and frequently using "public service" to refer to the achievement that had so moved them. One exuberant writer went so far as to declare that the *New Yorker* had "performed the greatest public service since the Crucifixion."[28] Another writer to the *New Yorker* clearly defined his understanding of the stakes: "It clearly called for editorial courage of a high order to devote so many columns of space, three weeks running, to this material, in the face of two risks: (1) the obvious one, of offending the chemical industry with its advertisements, and (2) that of depressing or boring such readers as might have found it all rather heavy going."[29] As time went on, the phrase "public service" likewise appeared repeatedly in letters to Houghton Mifflin and to the print media, along with words such as "courage" and "integrity." Meanwhile, those who criticized a publication's performance often wrote with a subtext that the publication had failed its public—by misusing facts on either side; by failing to present one of the two sides adequately; or by misrepresenting the journalistic vehicle (for instance, by presenting as an article a piece that the writer thought should have been identified as a book review or an editorial). And there were, of course, direct criticisms of the editorial position taken: "I protest the publication of your recent editorial in which you say an A&M official says we should pay no attention to Rachel Carson's book, *Silent Spring*."[30]

Those letter writers who criticized Carson (rather than the publication) in print did so in quite familiar terms. They questioned her accuracy or her use of facts, declared her "no scientist," or charged her with using irresponsible "scare tactics." While letters written directly to her publishers were sometimes not shy about accusing her of Communist sympathy or motives, those written for publication (or at least those screened by editors) were less likely to risk libel with outright accusations. A number took public issue with particular arguments she made, based on their own expertise or experience; and many referred readers or an offending editor or reviewer to the countering evidence. Often that countering evidence was the literature circulated by the organized opposition, such as Baldwin's *Science* review or *Time*'s panning article. For these writers, the support by association was framed in terms of the need to air and to shore up the "neglected" other side of the coin.

Pro-Carson letters in print frequently expressed their praise by way of defending her, often underscoring such points as her credentials and the documentation in the book. When the *Milwaukee Journal* gave front-page coverage to a speech by Stare, a letter to the editor summarized his attack and retorted in terms adopted from Carson's public pronouncements: "Rachel Carson, being a biologist and one with considerable field experience, is certainly well qualified to give an accurate account of pesticides and insecticides (or biocides as they should be called because they affect all living matter) and their relation to environment. . . . For those who seek further proof, she has listed in her bibliography about 50 pages of reference material."[31] Supporters sometimes urged other readers to consider the sources of criticism, occasionally with direct reference to the chemical industry or the government; and like her opposition, they often cited the positive regard of other publications and other media.

Consciousness of the fact of controversy was a strong component of audience response, and the comments of many were grounded in the familiar narratives of battle. More noteworthy, though, was the impulse on both sides to offer their own abbreviated recapitulation of that debate—in terms that were concise, simple, and persuasively "objective," even when the letter carried a point of view. In offering their own quick summary of the debate, the letter writers acted out what they expected of the media by providing what they believed was a defining digest of the debate for the rest of audience, irrespective of any thumbs-up or thumbs-down position. In that mimicry, consciously or not, they were acting within an assumption of continuity between the media and the audience. To paraphrase Walt Kelly's Pogo, "We have met the public and they is us."

Finally, it is essential to observe the deceptively obvious fact that the great majority of published letters took a stand; their writers were thereby debaters—participants in and contributors to an established debate. They wrote expecting the letters editors to welcome additional entries into the discussion. They wrote expecting the interest of other readers. And they wrote with a sense of both opportunity and responsibility for arguing toward clarity and/or persuasion.

Mobilization: From Judgment to Action

Although Carson's first objective in writing *Silent Spring* was to give the citizenry information, she was most gratified to hear about actions actually taken on the basis of that information. A substantial portion of those who wrote to her and to the media seemed to share the belief that, in a democracy, knowing about a wrong demanded acting to right it. But ideas varied widely about what the proper action might be, not to mention about where responsibility for acting might lie. Letters written to the media occasionally included evidence that the writers themselves had responded to the call for change or mobilization—for example, that they had stopped using DDT or that their community had passed legislation against spraying programs. Far more often, public letter writing represented an intermediary form of mobilization. Letter writers, particularly in the case of print media, were somewhat more often concerned about spreading the word than about effecting actual changes in policy or practice. And quite a bit more often, writers described actions they expected *others* to take rather than actions of their own.

A great many of the letters to the *New Yorker* urged that the magazine take action to extend the reach of the articles beyond its select audience, including ideas about circulating reprints, drawing media attention, and seeking republication in the *Reader's Digest*. Letters to both Houghton Mifflin and the *New Yorker* directed them to send copies to an assortment of people in power, from President Kennedy to legislators to "our park commissioner who just sprayed all the elm trees."[32] Many others were quite media-savvy, as was evident in letters suggesting specific publications for syndication or reprints. One example was a letter from the anti-Carson camp to *C&EN*: "The amateur gardener or average housewife is not likely to read *Chemical and Engineering News, Agricultural Research Service* or *Chemical Week. . . . Life, Post, New York Times*, and other widely-circulated publications should be approached to print the substance of Dr. Darby's review, even if a one-page advertisement has to be purchased to do it. Again let me emphasize that wider publicity of the scientific viewpoint . . . cannot be accomplished through the limited circulation of technical magazines and friendly persuasion by neighborhood scientists."[33] And one unique gesture was the spontaneous and unsolicited placement of an ad for *Silent Spring* in a November *New York Times Book Review* by an obscure and somewhat quixotic magazine, *Free Deeds*.[34]

Recommendations to the government were most likely to be constructive but vague calls for more study, or generalized and often diffuse demands that dangerous practices be stopped. Occasionally ideas were more specific, as in the case of an early letter to the *Washington Post* suggesting that "the President should set up a special commission to (a) recommend appropriate lines of research and (b) propose a national policy to control the use of chemical pesticides."[35] This letter preceded by a few days the *Post*'s editorial guess that a presidential panel would be established; both the letter's recommendation and the paper's guess were fulfilled. In areas most directly affected by pesticide use, of course, calls for action were often quite specific to local circumstances, such as calls for cessation of particular spraying programs or for the establishment of local education programs for pesticide users.

Recommendations directed to other readers at first had taken the form of calling their attention to the *New Yorker* articles, but once the book was available, directives to read it became common. One reader wrote to a Topeka, Kansas, paper: "If you are a parent or grandparent, or are altar-bound, you . . . have a moral and ethical obligation to your family to read Rachel Carson's *Silent Spring*."[36] Parallel directives came from Carson's critics, as in a letter to the *Lakeview (Conn.) Journal*: "Fair-minded people in regard to the recent discussions on spraying with pesticides, especially those who have been influenced by Rachel Carson's book *Silent Spring*, should read the other side of the argument. Recommended for this is the very lucid article, 'Biology,' [in] *Time* magazine."[37] Remarkably few letters could be found actually advocating specific changes in consumer behavior, although many called for increased education and awareness of hazards. One of the few published letters even referring to consumer behavior appeared in *Time* magazine following its negative report of Carson's message, and this letter described the writer's own action: "After reading installments of Miss Carson's book in a magazine last June, I was so struck with horror that I threw out all my insecticides and sprays."[38]

Personal reports of actions taken by letter-writing readers revealed most directly the impact of the book's message, yet few saw print. An abiding theme of those unpublished letters to the *New Yorker* was that of passing along the articles to others, from family and friends to government and media. A number reported their efforts to get media attention for *Silent Spring*, including writing letters to local editors and planning to review the book for local newspapers or radio stations. Several included lists of people to whom they had sent copies of the articles, including members of the media, legislators, and and government officials at all levels. Some offered to facilitate distribution ("I will contribute to a fund to circulate this worldwide");[39] and J. H. Rand, then retired founder of Remington Rand, reportedly contacted Houghton Mifflin wishing to buy "enough copies to stock public libraries in two thousand leading cities" in the United States, as well as to provide copies to all members of the House and Senate, "wives of the governors of the fifty states," and "wives of Chief Health Officers" of each state.[40] Eventually 2,700 copies were sold and shipped to a Bahamian research institute for his purposes.

Letter writers reported other answers to the "what can I do?" challenge that were more directly aimed at policymakers. There were promises to campaign for governmental action and reports of creation of local community action groups. Reference to such efforts appeared frequently in correspondence to government offices, notably in correspondence to the secretary of Agriculture. Although obviously not intended for public or media consumption, a large proportion of these letters had been addressed to members of Congress or the White House and referred to the USDA secretary's office, which was then charged with drafting responses within departmental public relations guidelines set up for the occasion. The USDA offices were much taxed to respond to the letter writers and were often required to send copies of their answers to citizens around to the White House or to Capitol Hill to verify that the original letter had been answered. A collection of stock responses was generated, but one weary assistant secretary responded tellingly to a crusading writer who promised she could collect many petitions from others who felt as she did. He added the following sentence at the beginning of the otherwise boilerplate three-page explanation of the department's position: "Dear [B.]: You do not need to send the petitions."[41] A number of letters came ready-typed with a line at the bottom for the sender's signature, obvious evidence of a letter-writing campaign at work. One letter writer made no effort to hide the fact that he had undertaken his own letter-writing campaign but rather added: "P.S. Please excuse the carbon. I am sending many of these."[42]

Awareness of media-focused activity was also a common thread in the correspondence to federal officials. At one point, USDA Secretary Freeman had told the media—downplaying the potential impact of *Silent Spring*—that he did not think the public was "receptive" to information about pesticide dangers. Annoyed at his sweeping assumption about American attitudes, one member of that public audience wrote him indignantly, "When Miss Carson's book came out, the public was quite definitely 'receptive.' "[43]

The actual number of letters sent to government offices cannot be ascertained, but indirect evidence suggests that they were received in unusual numbers in a rhythm paralleling the course of the debate in the media. Based on correspondence to the USDA and mentions in the *Congressional Record*, the intriguing impression is that citizens wrote to those at the federal level in greatest force immediately following the *CBS Reports* program in spring of 1963, rather than during the previous year after the summer *New Yorker* "ruckus" and September publication of the book. An obvious explanation is that many more saw the program than read the articles, the book, the reviews, or each other's letters. However, a number indicated that they had already read the book but were prompted by the program to write. The program had discussed the lack of interagency cooperation as a core problem in arriving at clear understanding of pesticide hazards. Therefore, even if the letter writers had not entirely grasped all the political intricacies, they likely drew the conclusion that the problem, and therefore the solution, lay at the federal level—in contrast to the regionalization or even localization that might have occurred within regional or local coverage of the debate.

Book versus Magazine: "This Should Be a Book"

Whether a citizen was so inspired by a television program that she wrote a senator, or a chemical company executive was so annoyed by a book review that he wrote a newspaper editor, the point of reference in these letters was ultimately *Silent Spring, the book.* Even in the summer before publication of the Houghton Mifflin volume, once it was known that *Silent Spring* was about to appear in book form, discussion was directed at it as a book, despite assumptions about its content made on the basis of the magazine articles. One might presume that letters to the *New Yorker*—written without expectation of publication and sent before it was widely known that *Silent Spring* would come out as a book—would reveal relatively little about audience response to the book form of *Silent Spring.* Nevertheless, beyond their usefulness as examples of public response to a media organ participating in the debate, they offer a surprising opportunity to discover what the book form—particularly in contrast to periodical publication— meant to *Silent Spring's* audience.

The magazine received a near-record number of letters, totaling 491, in response to its three-part series.[44] Nearly half were seeking the articles in reprint form, separate from the magazine. The most frequent reason given was the need for "wide dissemination" or "wider circulation" than the *New Yorker* afforded. At the time, the *New Yorker's* circulation was approximately 430,000, yet many writers referred to the readership as narrow and select—probably in part because of the *New Yorker's* intentional creation of exactly that impression. A typical comment was some variation on: "The *New Yorker* is a fine magazine, but this should be made available in book form so it reaches beyond the small group of subscribers." Yet few books of the era saw initial print runs of as many as 100,000; and indeed, Houghton Mifflin hurriedly had to revise its initial projected print run from 60,000 to 100,000 as the response to the *New Yorker* articles began to suggest the scope of the topic's appeal. It would be months before the number of books available from Houghton Mifflin surpassed the *New Yorker's* circulation.[45]

Yet the readers still expected, even demanded, that Carson's message be provided in book form. Nearly a quarter of those who wrote the *New Yorker* said outright that "this should be a book." In addition to worries about the limitations to a magazine's reach, the impermanence of the magazine format was a problem for some, noting that their pass-along copy had already become tattered or that they were worried it would be lost or not returned by a borrower. For those wanting to place the message in the hands of the powerful, the power of a bound book was desirable; in fact, one writer said that congressmen should all get reprints or copies of the book but that President Kennedy should get one "morocco bound."[46] A few were even concerned that readers, who should be "made to read" *Silent Spring*, would not stay with a magazine series because of the discontinuities of time and, conceivably, physical presence. For them, ownership of the arguments in book form would facilitate the actual reading of *Silent Spring*. A number urged that the book appear in paperback form, for considerations of cost, dissemination, and facilitation of ownership that outweighed the question of permanence. For others, the material needed to appear without the distraction or undermining effect of cartoons, light pieces, almanacs, and related ephemera—or of advertisements.

Almost none who wrote, however, either to the *New Yorker* or to any letters-to-the-editor sections, remarked on the possible significance of economic differences between books and other—advertising-supported—media. A few who touted Carson's courage against the opposition's well-financed "propaganda" campaign touched glancingly on the point: "While competent authorities have meager resources at hand in the plea for sensible use of pesticides, the vast arena of commercial advertising is open to the chemical industry."[47] The readership counted on but took for granted the probity of the book form, which depended on the independence of an unaffiliated book author and of publishers seemingly immune to industry or commercial pressure.

Finally, for readers in the summer of 1962, the possibility that the book might contain certain information not found in the *New Yorker* articles would have been just a matter of guesswork; and only one or two actually raised the question of documentation to support what appeared to them to have been substantial research. But by the time the audience had begun to read the book and write to newspapers, a number seized on the "fifty-five pages" as an important reason to accept Carson's arguments. Those who expressed the hope that the work could be condensed or digested, however, did so in the interests of extending distribution more than as a shortcut through the technical material. None suggested finding and reading the articles in preference to the book. For them, the book held the position of authority. And they wanted others to read it.

Again, the Unread Book

The charge of not having actually read the book was a recurrent theme in Carson's rebuttal to her critics; and among those members of her audience who supported her, actually reading the book was often an important point. But for others, like the Pennsylvania agricultural agents who had not read the book but who "disapproved of it heartily,"[48] reading the book was not essential. One woman writing a letter to the *Rochester (N.Y.) Democrat and Chronicle* editor was unabashed about using a book review to stand in for the book: "I have not had the sad pleasure of reading *Silent Spring*. . . . I recommend the excellent review of Miss Carson's book in the December issue of *Scientific American*. LaMont C. Cole, while not embracing all of Miss Carson's conclusions, agrees with her to an extent which should leave thoughtful and worried anyone who is not intimately connected with the big chemical firms or sold on the 100 percent efficiency of the U.S. Administration."[49] She completed her letter with a lengthy quotation from Cole's review. In contrast, another writer in Kansas, fully aware of such practices, exhorted other readers to "Get your information first-hand; don't read the 'reviews'; read THE BOOK!"[50] Whichever the sentiment, such writers demonstrated awareness that some of the *Silent Spring* debate was taking place outside of actual reading of the book.

Among members of the general media audience, it was understood that as a matter of time, convenience, and information sharing, some books would come to their attention that may or may not be read but whose content was important to know about. Therefore, the impulse to provide a precis of *Silent Spring* or a digest of the debate in letters to the editor involved the assumption that many more people could and should *know about* the book than would actually *read* it. In that assumption was the clear understanding that the "reach" of *Silent Spring* could be immeasurably broader than the already impressively large total of all copies of *Silent Spring* in print—including *New Yorker* copies, Houghton Mifflin's many print runs, Book-of-the-Month Club and Consumers Union editions, Fawcett paperbacks, newspaper serializations, and magazine excerpts and condensations.

Full-Circle and Shared Public Service

A virtually indefinite reach was what Carson had implicitly wanted for her message, even if the pro forma goal, shared with her publishers, was that as many people as possible actually read her words. The welcome news in the feedback that the audience's letters provided was that a population had, one way or another, "got it." And what they wrote in their letters provided a gratifying mirror of the terms of *Silent Spring*'s own mission. Even Carson's second- and third-order messages about the need for research, the relationship between research and industry, and the relationship of industry, government, research, and practice were to a greater or lesser extent also received, thanks in part to the energetic public efforts of her detractors. Nonetheless, for the opposition, too, the public's letters not only extended the reach of the rebuttal message but also provided testamentary reassurance that their counterarguments were being considered.

The hope that action would be taken as a result of "getting it" was also fulfilled, though not necessarily as Carson (or her opposition) might have expected, given the audience's preference for echoing and spreading the word rather than taking direct action. Yet this responding audience—whether responding to the message itself or to the media through which they received it—were much more than passive recipients of a mediated message or inert witnesses to a media-borne debate. The very fact that so many of the audience saw their role as extending the reach of the message *or* its refutation indicates a strong sense of active and public participation.

Silent Spring's audience justifiably assumed that the book had been written for them, as individuals but also as members of a public whose opinion and will were of interest to one another and to policymakers. Those who wrote to the media, whether for publication or not, did so out of the full expectation that they were invited to participate in the public debate—that their feedback, their evaluation, their reports of action, their recommendations were not only appropriate but, in fact, were the whole point of having publicized the debate about pesticide use. They undertook their communication actions within the inherent authority and responsibility of an informed public—acting much in the spirit of Rostand, whom Carson had quoted in *Silent Spring*. A letter to the *Bethlehem (Pa.) Globe-Times* declared: "I want to heartily applaud your stand of truly 'maintaining the people's right to know.' "[51]

That they assumed a functional position in the main nexus of the media system was demonstrated each time a writer drew on material found in another medium—even the book itself—and referred other readers to it. They acted, however, within the seemingly unstable balance between the vocal lone citizen—or even the "mute common man" for whom Carson was validating advocate and surrogate—and the citizens' collective authority, whose valued support Carson and her opponents courted. For the audience itself, however, that instability was resolved in an image of a public continuum in which they were as integral as any other part of the media system.

The audience's assumption that the media expected and even welcomed audience participation on democratic principle might have been naïve, given the nonpolitical benefits to the media in promoting a newsworthy squabble between the petite Miss Carson and the heavy-footed pesticide lobby. But the audience's expectation of inclusion in the debate is nonetheless the dovetailing correlative of Carson's own assumptions about the good that writing and publishing a book might do. Carson and her readers shared the belief that a book could effect a coalescence of political weight—not only through informing (the citizen's "right to know") and mobilizing but also by dint of its very independence from the political forces so needing reform. By extension, letters about that book operated in a similarly independent segment of the public sphere, at least to the extent permitted by the letters editors. Seigel's summary is apt: "The letter to the editor is more than idle epistolary chatter. It is the reader's lance as he tilts with City Hall. It is his passport to a community of interest with his fellow man, or an arena of controversy in which he can test his thinking with those who differ. It is his town meeting in print, his approach to a purer form of participatory democracy."[52]

Unlike reviewers and literary critics who concentrated on the literary or scientific qualities of *Silent Spring,* few of the readers' letters mentioned Carson's writing style or whether the subject matter was too technical (and those who did were uniformly other authors). Much more central to the popular response to the book was the ubiquitous theme of having performed a great public service. Especially telling were the many who wrote begging Carson or someone else to turn a similar investigatory and crusading spotlight on other social problems: "When will we get someone to write a 'Silent World' to wake up the conscience of people everywhere and the leaders of the nuclear powers to the dead end toward which mankind is being led by atomic testing and the ever-mounting stockpiles of mega-killers?"[53] As a writer to the *C&EN* explained, "In order to obtain a mass popular-type audience, the crusading type of journalism used in this book is necessary."[54]

When readers wrote to the *New Yorker* that "this ought to be a book," they acted out of a sense of their own duty, if not the *New Yorker's* as well, to see to it that *Silent Spring* had further airing. Second and just as important, they were saying that the book form was the medium best suited to performing the public service of bringing an important issue to attention of the public at large. As a book, *Silent Spring* had the best chance of being taken seriously, not just because it would be a physically compact, documented, and permanent record of Carson's message but at least as much because the media treatment of so important a book meant that its message had the best chance of reaching the largest number of people—regardless of whether all in the audience had read the actual text. In the view of such writing readers—and also thanks to the actions they took based on that view—spreading the word in the public forum worked better by means of a book than by a magazine or any other medium.

CONCLUSION

"Speaking Truth to Power"

Traditionally, *Silent Spring* is said to have caused an uproar, or to have prompted the DDT ban, or to have been responsible for creation of the Environmental Protection Agency. The leap from book to consequence takes for granted an intricate, intervening process of public attention and response. In the mid-twentieth century, that process occurred largely in the mass media, yet it remains the book *Silent Spring* that holds the historical position of importance. As the present work has recognized, of course, the warnings did not first reach public view in book form but rather from the *New Yorker* articles. Yet except for some brief moments early in the summer of 1962, no participant in the debate referred to the articles as the representative record of Carson's message. It was the book that stood for Carson's message, for better or worse, in the public eye. That phenomenon invites the question, why? What did it mean to author, publisher, opposition, and readers that Carson's warnings were communicated in book form?

If we use *Silent Spring* to uncover some sort of description of the book form, we see quickly that referring solely to the physical attributes of a book—paper, ink, front and back covers—makes about as much sense as defining television as a plastic chassis with circuit boards, speakers, and a screen. Rather, we should begin at the more abstract observation that a book is a means of communication, a reason for being that it shares with all media. From there we can note that it is a textual entity having certain dimensions, but those dimensions are better described in intellectual than in physical terms. Moreover, compared with other media, a book offers to both writer and reader considerably more information capacity—enough in the case of fiction to pursue a long or complicated narrative, and enough in nonfiction to explore a complicated subject. The *New Yorker* faced a considerable challenge in accommodating all of *Silent Spring*, even with Shawn's willingness to push the usual journalistic limits of space and time. The difference between the effect of the articles and the effect of the book resulted partly from the need of the magazine to provide shortcuts through the intellectual journey to cast the message in the more traditionally journalistic form, with drama first, defending facts and discussion later. Periodicals, furthermore, have a shorter half-life than books do because they are intended, usually, to be timely and by definition transient. What appears as a matter of news in a magazine or in a newspaper or on a news broadcast may be a brief "truth," with another supplanting truth arriving soon thereafter. The "truth" recorded in book form is presumed to be enduring, worth its shelf space because that truth will last as long or longer than the covers and pages on which it is printed.

Readers' exhortations that "this ought to be a book" were grounded in part in a perception that what *Silent Spring* said was a ponderous "truth" deserving of record in the most durable of print media. In addition, collecting all the text in one "compact" publication permitted readers to have all text and references in one place at one time, in relative informational quiet without the distractions of advertisements, cartoons, and other unrelated material. In the context of a burgeoning information age, those readers' responses could be described as seeking a book's unique ability to escape the "noise" common to other communication channels. Moreover, the very physicality of its permanence made the book a uniquely objectified form of communication, such that a reader might pass on its message by physically giving the book to someone else.

The compact and permanent weight of the book form of *Silent Spring* also had important correlatives in the institutional contributions to its public life. The term "value added" has been commonly used to refer to the value of editorial screening, filtering, revising, and vetting that takes place within a publishing house. For *Silent Spring*, Brooks's editorial involvement was, with Shawn's, only the first stage in a considerable expenditure of institutional resources to produce, promote, advocate for, and defend the "truth" of *Silent Spring*. By the time *Silent Spring* reached bookstores, the weight of its publishers' sponsorship had been added to that of its author's reputation and credibility. Whether or not the audience/readers accorded special status to the branded involvement of Houghton

Mifflin or the *New Yorker*, clearly the fact that some publishers took Carson's message seriously enough to print it invited readers to do the same. From there, it is a very short step to recognition that having the media take *Silent Spring* quite seriously in their turn was still more important to its audience. For *Silent Spring*'s opponents, that fact above all mobilized and determined their response.

The traditional view of the book as a mode of communication is that it is a particularly intimate medium between individual author and individual reader. Many would consider books the least public of all mass media, and indeed that impression may well be why books are routinely excluded from media studies. Yet when Rachel Carson embarked on her project, her intended audience was a more collective idea of the public and the media system as well. A point I have made repeatedly here is worth making once again: *for the single communicator, no other medium affords the same direct access to the entire media system.* Books provide that access partly, though only partly, because the media system is abidingly needful of content. Although content can come from anywhere, the content provided by books has substance and comes in compact and intellectually underwritten form. In the terminology of theory, we can say that books are able to set the media agenda because they can provide the media with an information subsidy for discussion of complex issues of public concern. In other, more concrete terms, a work like *Silent Spring* supplies the media with information that has already been researched, vetted, and packaged; and that information may well be about an issue the media has hitherto been unable to process for public consumption.

Books' ability to perform this function is much enhanced by perhaps their most significant trait compared with other media, that of independence. When Carson said that books were her "proper medium," it was because in writing a book she owed her authorial soul to no one. Had *Silent Spring* carried advertising for, say, an organic fungicide, all of the probity in Carson's credibility or Houghton Mifflin's imprimatur would have vanished. And even though the *New Yorker* supported itself through advertising sales, the perceived independence of articles like *Silent Spring* was preserved not only by the magazine's partition of editorial policy from advertising but also by the articles' parallel existence in book form. Carson's effectiveness as national town crier on the question of pesticide abuse hinged on the perception that her mission was solely in the public interest, that she wrote as a concerned private citizen to other citizens, in debt to no other interest in the public or commercial arena.

As the media system at large picked up the *Silent Spring* debate as a matter of news and discussion, the integrity and independence of what Oscar Gandy termed an "information subsidy"[1] that the book provided to the media was likewise crucial. Journalists would have been derelict had they failed to uncover some unholy relationship between Carson and an interest group, and her opposition worked hard to find just such a discrediting relationship. By the values of the day, it was laudable in a free-speech democracy for a book to carry a message of "defended truth," even to the point of calling for civic action. However, to be simultaneously giving comfort to an external interest, advocating commercial action (other than buying the book itself), was not. Other members of the media also relied on the disinterestedness of author, book, and publisher, as they reviewed the book, recommended it to others, or used it as stimulus for their own observations on life. Were that independence compromised, their own credibility would have been affected. In turn, they added their own credibility to the book's public life.

In this context, reference to the goals of democracy points further to an aspect of American culture that places special value in the lone voice, particularly when that voice "speaks truth to power" (to adopt a phrase from the early Quakers). The figure of Carson, petite crusader beholden to no employer and no sponsor, typified in many ways the image of the lone pamphleteer able to move an entire nation on the strength of personal knowledge and convictions. The reaction of chemical industry spokesmen served, as they may now understand, only to spotlight that figure, even as they made concerted efforts to discredit her or to redraw her image. Carson's "blunt instrument" (as illustrated in Gordon Brooks's cartoon for *Yankee*, which appears here as frontispiece) was her book, a weapon of direct and uncompromised communication. The respect accorded the book form is easily seen in media efforts to imitate its strongest attributes, as when CBS willingly forfeited advertising dollars and emphasized the long months of research for its program. Similarly, certain of the print media sought to duplicate the impact of the book's documentation by using the documentation in the "fact kits" provided by *Silent Spring*'s opposition, in order to join the debate at a more respectable level of authority than merely saying "she's crazy."

Eventually, however, participation of the media in the debate, *especially* when mounted in such a serious, authoritative manner, worked to relieve the public at large of actually having to read the book. As dismaying as it might be to publishers and bibliophiles, the reality is that Carson was probably less concerned about whether everyone in her audience read *Silent Spring* completely (or at all) than she was that those who discussed the book publicly had read it, and read it carefully. For the audience's part, although some private citizens wrote to the media urging others to read *Silent Spring*, many saw their role as offering a synopsis of the book, the opposition rebuttal, or both—or even excusing other readers from reading the book simply on their say-so. Moreover, a large portion of the letter writers were responding not to the book but to some aspect of media coverage of it.

Within the media, book reviewers must certainly understand that as a direct consequence of their summary and discussion, a book may not get read, even one they admire. The ability of a book review to stand in for the book and to fuel the engines of social discussion was described well albeit cynically by Gerald Howard: "They are marvelous guides to up-to-the-minute intellectual decor for people looking to appear *au courant* with minimum time and effort. Richard Rosen in *New York Magazine* coined the phrase 'bullcrit' to describe the use of reviews as cheat sheets for dinner and cocktail party chatter among the cosmopolitan set, for whom being caught with the wrong opinion (or worse, no opinion at all!) about the latest hot book *du jour* can have devastating social consequences."[2] From the point of view of such party-goers if not the population at large, the important thing is not necessarily that they read the book but that they know what it said, which means that someone somewhere must have read the book and told them about it. That "someone somewhere" is most likely to be a member of the media; and discernment comes in choosing the appropriate corner of the media to trust for one's purposes, whether social, economic, political, or even academic.

In sum, the mission and message of *Silent Spring* were thus cultivated both within and outside the covers of a physical book—a statement that should surprise few. The dynamics of its impact, however, were almost entirely functions of the media system into which the book emerged, dependent on the differential role played by the book form within that system. True enough, the modular relationship between author and reader was both the theoretical and the real core of what *Silent Spring* was meant to do: inform and mobilize. Numerous letters to Carson reported a private response or an individual action, and perhaps some correspondents did learn about the book solely by seeing it on a bookstore shelf. Far more often, though, the audience arrived at *Silent Spring*'s message by any of a wide assortment of media means. Once there, they may or may not have bought the book, let alone read it, in order to reach understanding of that message. Their understanding was furthermore almost certainly informed by a wide array of *others'* understandings as carried in the media channels—both explicitly in what was said about *Silent Spring* and implicitly in efforts to simulate the authority and independence of the book form.

It is absolutely obligatory to state as clearly as possible that almost everything said thus far depends more on sufficient perception than on necessary reality. For example, magazines bound for library use are far more durable than many, many books. Some newspaper articles are far more substantial and cognitively "weighty" than anything on the miles of shelves of humor, romance, homespun philosophy, and so forth. More significantly, the majority of book authors do not manage to gain the access to the media that Carson had, but their hope and expectation that they might is an important part of their choice of the book medium for their message. As to "truth," it is almost embarrassingly obvious that publication in book form does not guarantee anything about veracity, let alone clarity of vision. The ability of a work to command the respectful attention of the media depends on the perception that book authors, editors, and publishers possess certain qualities and have performed certain tasks inherent in publishing a book. *Silent Spring* worried its critics because of that perception and, further, because of their understanding that as a book it would attract media attention and generate credible debate. The ability of the media to conduct such a worthy debate on an issue depends, similarly, on the perception that participants will also hold themselves to certain standards of thought and behavior, even if they dismiss the premise of the originating book as *Silent Spring's* critics did.

Independence, too, can be a matter of perception but in a more complicated way. The difference between advocacy and persuasion can be more than semantic when placed in the context of public interest, as demonstrated by the charges of "propaganda" leveled by both sides in the *Silent Spring* debate. The word "propaganda" carried not only the freight of Cold War anticommunist sentiment but also the concomitant implication that the information was unreliable because it served hidden purposes, ones conflicting with the apparently altruistic public-spiritedness of the content. The charge did not stick as well to the book as it did to its opposition. The reason it did not was that the result of anyone's accepting and acting on Carson's warnings did not appear to benefit her economically or personally—or even to benefit whichever interest groups (e.g., the Audubon Society) Louis McLean and Ezra Taft Benson darkly suspected of subversive intent. The worst anyone could say about the author and her publisher was that they were "just" trying to sell books, an accusation that continues to be voiced against authors and publishers.

In contrast, the consequence of agreeing with Carson's detractors could be thought of in terms of quite obvious economic benefits to them—benefits that had nothing to do with their arguments about the dangers of allowing pests to overrun a world that lacked pesticides. Although the public at large was quite unlikely to think consciously about the "cui bono" aspects of the debate—who would benefit if Carson were right and who if chemical industry spokesmen were right—certainly their perceptions about the sincerity of the arguments were affected by the fact that in one case the proponent was a book author and in the other the proponent was a profit-driven commercial interest. The suggestion that public-relations ploys might be at work was, furthermore, a negative one, particularly in an era when the profit motive per se was being questioned, as were corporate and governmental manipulations of public opinion.

Such observations take us rather far into the essentially journalistic domain of what Scott Slovic calls "literary nonfiction designed to raise public consciousness,"[3] which is an important but limited segment of modern American book publishing, though one that comprises everything from self-help books to presidential memoirs and perhaps a few exposés in between. Certainly, in their quest to raise public awareness, authors of such works count on the expectations and interactions described here. Similarly, it is now an almost universal practice among journalists, especially those at the highest level of achievement in their own media, to turn to the book form to make their most meaningful and lasting contributions to history. One thinks of Robert Woodward and Carl Bernstein, Walter Cronkite, Tom Brokaw, and even cartoonists like "Herblock" and Gary Trudeau. Moreover, surely authors of fiction have also shared the same expectations about commanding a particular kind of respect for having published a book and, to a varying extent, about being noticed and taken seriously by the media system. They, too, protest if someone accuses them of writing "just to sell books," and their publishers will promote their work as a matter of pride in having the author as one of "their" authors. Above all, to the extent that the authors intend their fiction as social commentary, their mission is not so different from Carson's. Indeed, to see the congruence one need only think of the two works to which *Silent Spring* is most often compared—*Uncle Tom's Cabin* and *The Jungle*.

Could It Happen Again?

The history of *Silent Spring*'s public life, however, does provoke speculation about whether it was a unique episode or whether the same thing could happen again. A very simple answer is that it did happen again, with variations of scale and type, as in the cases of *Unsafe at Any Speed*, *The Feminine Mystique*, or more recently, *The Road Less Traveled*, *The Bell Curve*, or even *Men Are from Mars, Women Are from Venus*, as well as a rising number of book-bound political polemics—to name just a few examples of news-making "nonfiction designed to raise public awareness." Were it simply a matter of making news, any number of controversial works of fiction could be added to the list: along with *Uncle Tom's Cabin* and *The Jungle*, such diverse examples as *Lady Chatterley's Lover*, *Valley of the Dolls*, *The Gulag Archipelago*, *Jurassic Park*, *The Color Purple*, *Bonfire of the Vanities*, "Anonymous's" *Primary Colors*, or even any of the Harry Potter series. Some works of fiction make news simply because they are intrinsically sensational and not intended to change policy or legislation, although social and emotional mores might be at issue. But even then, the media seize the opportunity for public debate on the nature of the sensation itself; and sometimes policy and legislation can be affected indirectly. In some respects, fiction even more than nonfiction has cherished the lore of the lone voice speaking from the pages of a book, unencumbered by obligation to anyone save a canny but sympathetic editor.

Still, *Silent Spring* does represent something of a cultural "perfect storm," in that it enjoyed a confluence of circumstances that made the episode a landmark not only in environmental history but in book history as well. Certainly the moment in history was especially welcoming for a work designed to stimulate civic response to an environmental danger from misused science. Precursory worries about radiation and the memorable 1959 cranberry scare arose in a youthful postwar society of increased wealth and education, giving rise to idealistic antibusiness rumblings, political activism, and general social criticism. It was also a society still relatively new to the elaborate system of information media that had added networks of radio, film, and most recently television to the existing print media. Particularly in the case of television news and documentary, experimentation with viable forms of public education and information was being conducted before the eyes of a public still inclined to consider such sources universally reliable.

Silent Spring also arrived on the transitional cusp between the now-somewhat-mythic "golden age" of publishing and the latter-twentieth-century industrial restructuring that ejected book publishing from an idyllic (and again, mythical) insulation from "bottom-line thinking" and relative isolation from other media industries. For itself, the work *Silent Spring* benefited from the devoted networks of like-minded people committed to speaking its "truth to power"—beginning with Carson's editor, Paul Brooks, and her agent, Marie Rodell, and eventually extending to people well-positioned to spread the word of the book and even to affect policy. Furthermore, Carson was already a popular, credible author of works politically neutral but authoritative, at least as far as her established readership was concerned. As for perceived independence, some disinterestedness seemed to be required of the media covering the debate; but pesticides were not a product requiring heavy advertising, nor were their producers heavy advertisers in the consumer-driven media of the day. The general media were therefore not often in serious jeopardy of goring the ox of someone on whom they depended for income, although they did risk compromised reputation if *Silent Spring*'s opposition could successfully charge them with bias. At worst, that simply meant that the media had a controversy on their hands demanding two-sided discussion, which was hardly a negative from the journalistic point of view.

Finally, the fact that *Silent Spring* has never gone out of print implies, first, that the work had some exceptional quality raising it above the status of a sensational but passing phenomenon. Had Carson's warnings been mere hallucination and easily debunked, the media might still have been able to generate a flashy public maelstrom, but the squall would have passed quickly. So it is important to acknowledge that in the esteem of many, her book is a classic of environmentalist writing—at once accessible and intellectually defensible. Further, though some of the science may be out-of-date, in the view of many people, the issues she identified are still very much with us, more demanding of resolution than ever. The endurance of debate over human intervention in nature and the endurance of the book *Silent Spring* thereby do indeed seem intertwined, for the book remains emblematic for both sides of the argument—locus of blame or credit, or simply a "landmark," as it is so frequently called. Without going any further into the debate itself, one can yet ask whether the book would still be in print had it not been written so well; whether it would still be in print had the problem of pesticidal hazards been miraculously resolved; but above all, whether it would still be in print had it not been marked so publicly as the revelation of an important public matter. And so we come back to the relationship between arguments made in book form and their treatment in the media system.

Silent Spring was ultimately the ideal subject for a study of relations between books and other media because it did manage to demonstrate so many aspects of a book's public life and its interactions with the other mass communication media. The insights and perspectives of everyone involved in the *Silent Spring* debate tell us much about what, to them, the book form represented and what it could do: what an author wanted it to do, what a publisher wanted it to do, what its opposition feared it might do, what the media depended on it to do, and what the audience thought it did or should do. In all cases, the dynamics between the book and the media system were essential parts of the expectation, not only as a matter of distinguishing a book from another medium but also as a matter of the shared purposes of books and the media. Thus, deciding whether "another" *Silent Spring* could happen again seems to depend on what we will expect books to do in a future much altered by changes in information delivery and perhaps even in purpose.

Implications for the Future of Books

Some of the intrinsic attributes of the book form may be unchanged whether the text is preserved on paper or in e-text; these attributes include the cognitive dimensions and the compactness of the presentation of thought, analysis, or narrative. A book is a communication of thought that is of such-and-such size, of such-and-such complexity, located in one piece in such-and-such a place, whether or not it is serialized in a magazine or downloaded piecemeal to a printer.

Permanence is more troublesome. The very essence of e-text is its mutability, not to mention its potential for nonlinear communication. Accurate or not, the traditional perception of the book form is that once a message is in print, it is fixed. Although readers may skim the work on their own time and in their own order, there is nonetheless an intended and expected progression of thought or revelation from beginning to end. By contrast, e-text enthusiasts love to celebrate the "democratic" opportunities for readers to assume the author's or editor's role in refashioning the narrative, as well as the creative possibilities of hyperlinks that provide author and reader with an endless network of cognitive routes and destinations. Even granting some elasticity in such definitions of what a book may be, the literal question of the physical medium still remains a basic one. A book may be recorded on a read-only CD or even in cyber-perpetuity on a secured website, but without an equally durable e-text reading device (not to mention a reliable power source), it becomes unreadable. Thus the question of the endurance of books as a physical medium entails more than the ergonomic question of whether "only Twinkie-charged insomniac dweebs like to read on the screen."[4]

With technological changes in book publishing, the observation that "anyone can write a book" suggests some erosion of what we once relied on as the special probity supplied by acceptance for publication. The removal of that editorial "value added" concerns those who worry on the public's behalf about the credibility of the book form. Efforts to replicate it among new media by subscription and other controls on electronic publishing suggest that editorial contributions to the book form are something we still expect or maybe hope will continue to be valued. Meanwhile, however, the political and legislative maelstroms around copyright and intellectual property issues have potentially enormous significance. The severe resistance of some publishers (including Condé Nast as conservators of the *New Yorker* archives) to having any text placed in electronic form reflects above all profound aversion to the nearly total impermanence and mutability of electronic text. One can only speculate about which force, among many, will prevail: pro- or anti-electronic text factions, authors, copyright owners, scholars, publishing houses, parent companies of any of these, or the reading audience's own needs.

Meanwhile, the endurance of books as a cultural medium is linked to their function in the larger media system, both differentially and dynamically. Within that system, the lore of the lone voice of dissent draws on the compelling, traditional American presumption of a public sphere into which that voice may speak and reach the full polity. The *Silent Spring* episode is an exquisite example of the special access to that public sphere afforded by the uniquely commanding book-to-media springboard; and it is also a model of the single voice magnified, multiplied, and ramified by the media. The growing number of preelection books by presidential candidates is a good contemporary demonstration of the idea that one voice can evade the barriers and requirements of other media to present their case directly to the electorate—at length and at leisure—while still taking advantage of the media's enormous reach. By contrast, an e-book published on-line in bytes and pixels at present has nowhere near the same potential as a "deadtree" book for entry into national discussion, despite the fact that an e-book can reach individual members of the electorate directly and instantly. The current cyberspace version of lone pampleteering seems to be the entropic and wholly unauthoritative phenomenon of spam, which fails to create anything like a civic locus for public debate—in fact, it creates quite a chaotic opposite.

Particularly on matters of public policy, public-relations agencies and corporate public-relations offices have been charter experts at utilizing the same media dynamics to control as well as to magnify the corporate or institutional voice, by the very definition of their work. Habermas, whose concept of the public sphere has been used here, particularly abhorred the practice and field of public relations, increasingly as time went on. He believed that a public sphere in which the citizenry determines policy through "rational-critical" political discourse is invalidated when controlled by agents of the private sector.[5] And in American legal history not long after *Silent Spring*, the courts came to deem the corporate voice as having the same rights and protections as the single citizen's voice, regardless of wide differences in resources to gain public attention.[6] Aside from some with political party assignments, public-relations professionals have not resorted largely to books to make a direct case for one industry cause or another, for mostly practical reasons—although one government agency, NASA, recently tried to use the book form to refute persistent charges that the 1969 moon landing was a hoax. Nonetheless, lines between promotion of commercial interests and book creation and publication are blurring as practices such as product placement cross over from other media. Certain authors specify brand-name jewelry, cars, drinks, and the like for their fictional characters; and in-house production of consumer-market books by conglomerates with "product synergies" such as movies, toys, and games is now quite common.

The question then becomes whether voices that are independent because they are unsponsored will still have access to the public sphere. That question has been raised often, usually with an implied negative answer, bemoaning the dissident author's plight in being unable to get his or her word to a PR-blinded public world. Similarly, potentially iconoclastic poets and fiction writers are often depicted as a threatened breed with vanishing access to publication and promotional resources. The pessimistic ones quote A. J. Liebling: "Freedom of the press is guaranteed only to those who own one." Authors may find some comfort, however, in the continuing (if threatened) existence of independent and university presses, as well as in the media's rather indiscriminate taste for controversy and sensation. One wonders, though, if the truly critical question may not be whether the *public's* reliance on the perceived independence of the book form will endure. Is that autonomy—that intellectual sovereignty which seems so fundamental to the American image of itself—something the public will continue to value and expect of some corner of the information system, at least? Our tolerance for the distractions of ubiquitous advertising, including ads that currently crowd cybertext into a small part of a website, seems to be limitless, as does our willingness to have any point of view be identified with any commercial interest. The very nature of independence seems increasingly ill-defined. Without debating the rightness or wrongness of this situation, we must ponder not only individual citizens' inclination and ability to distinguish advocacy from persuasion but also their a priori attitude toward anything they know—or think they know—came from a book.

Peculiarly, worries about real or perceived independence lead us back to the phenomenon by which books like *Silent Spring* become influential but widely unread. It was because Americans of 1962 received their information through multiple media that they understood the relative functions and status of each part of the media system. They had become accustomed to getting immediacy and visuality about a news story from television coverage while getting details and interpretation from a newspaper. They expected that what appeared on an editorial page would be opinion but that what appeared on the front page would not. Like generations of readers before them, they were accustomed to expecting book reviewers to be literate souls of some intellectual virtue, and they expected reporters on issues like pesticides to pursue "the facts" and "both sides of the story." The reality, however, might always be that certain book reviewers lack insight, thoroughness, or good will; and the expertise, honesty, and objectivity of journalists has been questioned on more than one occasion. The point, though, is once again less about reality and more about the expectation that the media choose worthy books and discuss them in a worthy way. At some level, the audience may even comprehend the media's continuing need for books' subject matter, insofar as the nonentertainment function of the media is to bring matters of public interest to public attention; and books contain someone's careful consideration of an issue, even if that issue is nothing more political than easing personal depression.

By this implied logic, the public permits itself to depend on the existence of books but, further, to depend on the media to screen and digest those books, concentrating the information along with its implications and shortcomings into quickly apprehended format and/or ongoing debate. This does not mean that audiences for a book will never buy it and read it; publishers and authors will still expect that media exposure will lead to sales and reading. For a large segment of the information-overloaded population, however, books may represent an important, needed, well-respected source of information—but information for some-one else to read, someone in the media reliably credentialed as a surrogate reader. On the basis of what is read in a *Chicago Tribune* book column or the *Newsweek* book page, or heard on *Fresh Air, Booknotes,* or *Oprah,* many more people will discuss a book than will actually read it.

This does not mean that an author's mission is unsuccessful. On the contrary, one has only to think of some of the most news-making works in recent history to find some of the least read—*The Pentagon Papers* and *Satanic Verses,* to name just two. The purpose of Daniel Ellsberg and the *New York Times* in making the papers available in book form was clearly not to make sure that each document would be read by every reader but rather to set in motion the inevitable media processes and, eventually, to affect policy. Salman Rushdie's artistic purpose in writing the *Satanic Verses* may or may not have had an overtly political component, but the actions of his critics demonstrated the same fear and antagonism toward his book-borne communication that Carson's critics felt—though raised to a far more deadly order of magnitude. Their response, like Carson's opponents', only served to invigorate the book's public life.

If the book form is to survive, then, it must be because society needs it for some purpose or another. At present, authors seem to need books as a durable way to record ideas or narratives of a certain complexity and length, and authors also seem to expect that writing a book may be the best if not the only way to get the broadest possible airing of their ideas via the media. Publishing companies seem to need books not merely as vendible commodities but also as a business activity of some prestige, one that may also provide synergistic benefit if a parent company needs intel-lectual property for another medium, such as a movie or a magazine syn-dication. The media seem to need books and authors for things to talk about. Those "things" may be anything from children's fantasy fiction with allegedly satanic overtones to exposés about pharmaceutical company mis-deeds. Those authors may be flamboyant politicians or single mothers scratching a first draft onto a paper napkin.

And what about audiences? Readers? Nonreading citizens? Audiences seem to need books for any number of purposes, including entertainment, self-education, gift-giving that sometimes includes indirect message-sending, and focal topics for social exchange. The people's "right to know" is an important justification and driving force for American media machinery; books have traditionally had a special status as source for whatever it is that people should know. Ownership of information production and distribution complicates matters to the extent that "the people" have traditionally had the expectation of finding "objective," unencumbered information somewhere in the media system; and books have historically occupied a privileged position in that regard. The relativism of the latter twentieth century, however, began to undermine the idea of objectivity in favor of the kind of "balance" that Carson's critics felt she lacked. Moreover, the function of dissent in American society may be undergoing a qualitative change—one that some have charged is closely related to the rise of conglomerate media ownership. One may argue that in the future no books of true dissent are likely to get discussed or even published in our increasingly-concentrated system of media ownership, but that argument fails in the face of the continuing supply of such books. In one recent example, Gerald Markowitz's and David Rosner's *Deceit and Denial*, the authors charge the chemical and lead industries with more of the same abuses of public trust that concerned Carson forty years ago.[7]

An ultimate question is, of course, whether the people's right to know necessarily entails the people's obligation to pay attention. The challenge of our contemporary "information age" begins with how much there is to know and how many ways there are to know it. The book form is viewed as a flag that a particular message has warranted particular attention, and the media's treatment of the book form has become an indicator that the issues merit special public interest. *Silent Spring* was a successful enterprise for Rachel Carson, the *New Yorker*, and Houghton Mifflin because attention was indeed paid, and action was taken. The experience of *Silent Spring* was an exemplar for future lone pamphleteers, but it also was an object lesson for those whom Carson took on, those in power, and those with vested interests.

The dynamism of a system of systems is that one subsystem survives by learning adaptations to meet threatening changes from the other subsystems. In the media system, the professional descendants of the chemical industry public-relations agents have taken to heart the lessons of avoiding engagement in controversy or of going on the positive offensive without direct reference to an original challenge. At the same time, however, books are being written about that very process, chronicling the dangers of public-relations sophistication when employed by the privileged and powerful, whether in the corporate or government sector.

For books as we know them to disappear altogether, a number of things would have to occur. First, and as unlikely as it still seems given the number of trees that have been sacrificed for printer paper, readers would have to prefer reading on-screen and relying on reading-device-dependent media. Readers would, moreover, have to be willing to abandon reading long, complicated works, and there is certainly evidence that that may be happening, as Sven Birkerts has observed and lamented.[8] Authors would have to have faith in the ability of some other medium to command the same respect and attention that books do, or else they would have to abandon the idea of reaching the public sphere as an integrated entity. For some authors, it is already quite acceptable to pursue communication with the "wired" readership on a one-by-one basis, which is seemingly little different from the original author-reader dyad of the pre-mass-media eras. However, one of the most lively activities on the Internet nowadays is that of reviewing things—products, services, software, hardware, games, and certainly books—for other people. Commonly touted as a quintessentially democratic phenomenon, the practice still suggests that readers will continue to rely on someone else to find and explain the worthy work, which leaves plenty of room for the persistence of the weightier tome as long as someone (else) has the urge to do the reviewing.

At the same time, however, the media system would have to change so completely that the intellectual capital reposing in books would lose its perceived value and the probity inherent in the book form would lose its cultural significance. Such a change could occur with the loss of perceived independence or with the devaluation of independence, and there are pressures toward both. Nonetheless, the multiplicity of what pours forth from our publishing houses seems to allow for many gradations—from the drugstore-paperback novelization of a science fiction film to the ponderously footnoted monograph of a university press. The advent of each new work of newsmaking social criticism or dissent, however, can provide a new case study and a new indicator of the social standing of the book form. If we do hear about "another" *Uncle Tom's Cabin* or *Jungle* or *Silent Spring* that speaks truth to power, it may be a fair signal that the book form is still valued and may yet survive.

APPENDIX

Perspectives on the Study of *Silent Spring*

Studying a book's public life meant drawing on the perspectives and methods of the several disciplines out of which book history has lately emerged as a field in its own right. To evaluate how *Silent Spring* and the media interacted, not to mention the significance of the fact that Rachel Carson's message came in book form, I found myself framing questions and strategizing how to find the answers in ways that constantly crossed disciplinary lines. In this appendix, I begin by noting the contributions of those who have written specifically on Carson and *Silent Spring* before me. Thereafter, I discuss some of the relevant philosophies and perspectives as they informed my work, organized insofar as possible along disciplinary lines despite my insistence that useful analysis required a thoroughly cross-disciplinary approach.

Many authors have already written about *Silent Spring* and Rachel Carson, including Carson herself. Her own history of the era is found in uniquely intimate form in the correspondence with her close friend Dorothy Freeman. Edited by Freeman's granddaughter, *Always, Rachel: The Letters of Rachel Carson and Dorothy Freeman, 1952–1964*[1] provides invaluable insights into Carson's struggles, attitudes, priorities, and work philosophy, as well as some details of her interactions with publishers, media, and opposition that are not found elsewhere. From the point of view of history, the biographical memoirs of Carson written by her close friend Houghton Mifflin editor in chief Paul Brooks are especially valuable. His quite personal biography of Carson, *The House of Life: Rachel Carson at Work*, provides considerable historical detail about the publication of *Silent*

Spring and Carson's life at the time. At least as important for this study, however, it is remarkably rich in Brooks's own experiences and perspectives; and it thereby constituted a primary research resource almost in its entirety for what it revealed about an editor-publisher's perspective on a controversial book. His memoir about Houghton Mifflin, *Two Park Street*, which concluded with a chapter on the publication of *Silent Spring*, was also useful.[2] During and after his long tenure at Houghton Mifflin, Brooks wrote several naturalist works of his own, including his 1980 *Speaking for Nature*, a history of American literary naturalists from Thoreau to Carson.[3] In that book, one quote from John Sears is of particular note, given Brooks's role concerning *Silent Spring*: "The ecologist, with all of his professional training, should be chosen with some regard for his talents as a publicist."[4] Brooks agreed to three interviews with me in the last years of his life, and he died just as I was writing the chapter on Houghton Mifflin. My experience with him and with his writings afforded an exceptional view of the world of publishing in his era, and I only wish there had been more time to talk with him.

Otherwise and much more recently, by far the definitive biography is Linda Lear's 1999 *Rachel Carson: Witness for Nature*.[5] An environmentalist historian herself, Lear offers an exhaustively thorough portrait of Carson's life and experiences, as well as special understanding of the significance of *Silent Spring*'s arguments and the response to it in both biographical and historical terms. Her unique access to certain people and resources, moreover, provided a few otherwise unavailable details. Several other biographies of Rachel Carson exist, including Patricia Hynes's *The Recurrent Silent Spring*, a particularly thoughtful and analytical treatise on Carson, *Silent Spring*, and the establishment response to it, with a well-considered exploration of gender issues.[6]

However essential Carson and her life's work were to this story, my concern was with the participants in the entire process of production (including author), dissemination, and reception of *Silent Spring*—their respective experiences and how they *each* saw the role of this news-making book. The most systematic approach to the discussion suggested itself in Robert Darnton's landmark "Communications Circuit," which provided the organizational framework for this study, with slight modification. His circuit traces the relationship and flow from author to editor and publisher to producers to distributors to readers, circling back to author (for "authors are readers themselves").[7] As my focus shifted chapter by chapter through the circuit, from one group to the next, I found myself turning to various

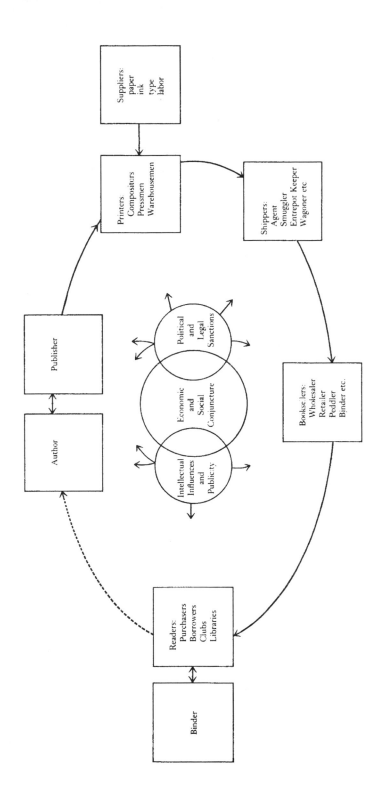

Robert Darnton's Communications Circuit. With thanks to Robert Darnton.

disciplines for a viable approach to the material. What follows is a brief survey of what seemed to be the most pertinent resources and concepts from several fields, with the necessary caveat (or confession) that the price of such a multidisciplinary approach is necessarily lack of depth in any one area. Writings from book history and journalism history intersected with those from environmentalism, mass communication, critical studies, and even public relations.

Logically enough, the specific work, *Silent Spring*, has enjoyed extensive attention from other environmentalist writers, the vast majority of whom explore the significance of the work in the history of environmentalism.[8] These writings tend to take the role of the media so thoroughly for granted that they often conflate the book with the controversy as stimulus for change, concentrating primarily on the resultant governmental or corporate actions. Few consider the media's role as anything other than intermediary or simply a historical given—the debate happened, the media were active—as proof of the book's power.[9] The conventional scenario brackets the episode of controversy such that, as James Whorton put it, there was a "BC"—"before Carson"—era and an "AC"—"after Carson"—era. His work is one of two nearly contemporaneous works that locate the episode firmly and literally between antecedents and legacy: Whorton's 1974 *Before Silent Spring*[10] and Frank Graham's 1970 *Since Silent Spring*.[11] In addition, Graham's work gives considerable behind-the-scenes information about policies and policy makers, and his personal experience with some of the actors on both sides rendered his book useful as a primary source. As a journalist himself, moreover, Graham had special awareness of the dynamics of the debate carried on within the media.

Those concerned with the rhetorical aspects of Carson's book and the debate around it more nearly approach dissection of the debate itself, but by definition their analysis is limited to the terms and themes used by partisans and observers.[12] Craig Waddell's recent anthology, *And No Birds Sing*, with a foreword by Brooks (his last published words) and an afterword by Lear, is the most focused study of the rhetorical eddies and whirlpools in the *Silent Spring* debate.[13] As such, the collected writings form a good exegesis and consideration of the controversy's content but less so its process.

Although the functional relationship between books and the media is rarely addressed in any depth in any discipline, a few writers on *Silent Spring* have examined certain aspects and limited dynamics of the media debate around that book. Valerie Jan Gunter used *New York Times* coverage of the *Silent Spring* debate to test a hypothesis that a "symbiotic relationship with news media" gives government "an inordinate amount of influence over environmental controversies" (the hypothesis was not supported).[14] Sociologists Andrew Jamison and Ron Eyerman cast Carson as foremost of a new breed of ecological-intellectual "counter-experts," who used arguments from science rather than politics to advocate political change. They wrote: "Nature writing and science journalism took on a new significance and could reach wider audiences through new production techniques and media as well as through the new methods of dissemination and communication. . . . What made Rachel Carson's story particularly important, however, was what she managed to do with her success and with the new contextual opportunities: she used them as a platform for social criticism."[15] What Jamison and Eyerman overlooked, however, was that Carson's platform was, to begin with, the primordial medium of the book.

One study that did offer a brief but excellent summary of actual functions of the media in the debate was Christopher Bosso's study of "the life cycle of a public issue," which identified three factors contributing to Carson's "impact on the national consciousness." First was the timing relative to the thalidomide tragedy, and second was the new, sympathetic receptivity to Carson's mission among the Kennedy administration and Congress. In describing the third factor, Bosso succinctly set forth one of the essential dynamics to which this study was devoted:

> Third, and probably most important in the long run, Carson's original writings were quoted, discussed, and publicized by other media on a scope not previously experienced for an "environmental" matter. That initial *New Yorker* series reached a limited audience, but particularly explosive pieces of journalism tend to have marked multiplier effects—a pattern of attention first on the part of political leaders and the "attentive public," and later the mass audience, as other media organs report on the original story. The first two factors influenced short-term government response; the third helped *Silent Spring* to spark the modern environmental movement.[16]

Particularly noteworthy in that light given the perceived missions of author, publishers, and supporters, Scott Slovic wrote about the phenomenon by which awareness is assumed, not always correctly, to lead to action. "Carson [was] concerned with the particular type of awareness known as 'public awareness,' the collective elevation of consciousness that is considered valuable mainly because it suggests the potential for political influence. *Silent Spring*, in fact, is the classic example of literary nonfiction designed to raise public consciousness."[17] Writing nonfiction to inform and raise public consciousness locates the work of such books squarely in the same tradition as periodical journalism, yet journalism history in the main has ignored the world of book publishing, including the career of such exposés as *Silent Spring.*

Unfortunately, the new and burgeoning field of book history, too, has yet to look very extensively at (Slovic's) "literary nonfiction designed to raise public consciousness," particularly that written after the explosion of mass media in the twentieth century. The field of history of the book arose out of literary history, antiquarian, and bibliographic studies to become what Darnton, a leading American scholar in the area, has called a "riot [of] interdisciplinarity."[18] Nonetheless, its literature largely reflects its roots, in that the majority deals with the material, social, and historical contexts of "serious" literary works and is often limited to books published from approximately the Gutenberg era through the late nineteenth century. The role of books in public debate has been considered, but rarely with respect to works of contemporary nonfiction, rarely from a sociopolitical vantage point, and very rarely in relation to other media.

Darnton is himself among those calling for greater diversity in approaches to book history, although his circuit (strikingly familiar to students of mass communication theory) did not in its published form directly address promotional activities or links to other media. Rather, it relegated them to the vague inner territory as "intellectual influences and publicity, economic and social conjuncture, political and legal sanctions."[19] Darnton's circuit did at least subsume the social, political, and economic circumstances of publishing—according to John Sutherland in 1988, areas neglected by much of "literary sociology" as part of the "machineries and the material processes by which books ('literature') are produced, reproduced, distributed, marketed, merchandised, and consumed."[20] That neglect is mitigated somewhat by Elizabeth Eisenstein's two-volume opus, *The Printing Revolution in Early Modern Europe,*[21] on the complex and far-reaching social changes wrought by the advent of the printing press, and by the countering study of print culture in the same era by Adrian Johns in *The Nature of the Book.*[22] The recent publication of *Perspectives on American Book History*[23] has added a valuable and specifically American set of historical studies of literacy and print culture. Currently in progress at Cambridge University Press is a five-volume series, *History of the Book in America,* which promises to focus more directly on cultural issues, including readership and reading. Moreover, the institutional history of publishing has enjoyed a steady stream of corporate histories of publishing houses and memoirs of textophilic editors and publishers, which has only recently begun to dry up for the very relevant reason of major structural changes in publishing.

Generally speaking, there have been two waves of writing on the American book-publishing industry, one in the 1960s and one in the mid- to latter 1980s in the era of mergers and conglomeratizations. The first wave peaked at roughly the same time Houghton Mifflin was publishing *Silent Spring*. It included the writings of a generation of editors and publishers as they confronted major postwar changes in the structure and culture of what they had known in publishing's prewar "golden age"—often as they faced their own retirement. John Tebbel's commanding four-volume history of American publishing provides a substantial record of some of the facts and dynasties of the industry up to the end of the 1970s. Oddly, although a media historian himself, Tebbel never drew the two worlds together in any meaningful way.[24] Institutional examinations of publishing have generally skimped on cultural context, but notable exceptions are Hellmut Lehmann-Haupt's thoughtful history of American publishing, written in 1951 and not really paralleled since, and the 1982 study in *Books: The Culture and Commerce of Publishing* by Lewis Coser, Charles Kadushin, and Walter Powell.[25]

With the exception of O. H. Cheney's landmark 1931 study,[26] studies of American publishing's economics outside these publishers' memoirs were otherwise not common until the panicky second wave of writings in the 1980s. Yet even at the time of *Silent Spring*, the field of business management was rapidly becoming professionalized with an attendant rise in publications in the "science" and theory of business. Among the pivotal concepts in management science was that of *risk*; and book publishers of the era already understood that a defining trait of their industry was a form of risk-taking different from, or in addition to, purely fiduciary risk. Pierre Bourdieu's conceptualization of risk suggests a useful framework within which to understand publishing's brand of risk-taking, insofar as Bourdieu posited an inverse relationship between economic and cultural capital and, relatedly, economic and cultural risk. In his model, small producers of cultural products tend to enjoy more cultural than economic capital, such that they are able to take greater risks in cultural innovation than large producers, who lose cultural capital as their economic capital increases.

For publishing at large, the twentieth century's challenge was to sustain the preferred balance between cultural and economic risk-taking in the face of profound structural changes. At the time of *Silent Spring*, the paradigmatic approach of a successful publisher was to find and use best-sellers to subsidize risks taken on worthy but potentially unpopular books. However, cultural and economic risks sometimes needed to be assessed in light of such realities as the political implications of a public debate. In the case of a controversial book like *Silent Spring*, many forms of risk were taken by many involved in its creation, promotion, and even in its rebuttal. For example, the opposition's risk in failing to respond publicly to *Silent Spring* ranged from immediate loss of pesticide sales revenue to longer-term political actions that could affect procedures as well as profits; the opposition occasionally even framed its alarm in terms of harm to public belief in industry integrity. Hindsight now suggests that shifts have oc-curred in the definition of risk since *Silent Spring*, and the practitioners—the executives, the managers, and the public relations officers—may pro-vide the best window into that issue.

Indeed, some of the most useful works on the culture, mechanics, and economics of publishing at mid-century came from the practitioners themselves rather than from the academic world.[27] Yet, with the exception of a few "how-to" references,[28] relatively little specifically concerning pro-motion and publicity was written, perhaps out of squeamishness about considering book publishing a business rather than a profession. Beyond what was written in public-relations and trade journals such as *Publishers' Weekly*, most of my understanding about the general nuts and bolts of sixties-era book promotion came from the archived Houghton Mifflin correspondence and materials, and from the memoirs of those involved in *Silent Spring*.

At the fulcrum between a book's publishing history and its public life, however, is the entire realm of book reviewing, about which a fair amount has been written in the scholarly world. However, for practitioners, review-ing is part of promotion, while for book history scholars, reviewing is part of literary critical studies—and never the twain shall meet, it seems. This traditional cleavage derives from a distinction between criticism and jour-nalism made in theory but largely untenable in book-reviewing practice. The scholarly image of book reviewing as an exercise in criticism typically concerns fiction and "serious" literature and focuses essentially on the artistic or intellectual product of the author. Cross-review debate may arise but usually in the form of discourse about the book's literary worth or historical accuracy. Often overlooked and of critical importance regarding

Slovic's "nonfiction designed to raise public consciousness" is the role the reviews play with respect to each other, to the media in which they appear, and to the public discourse around the book in question. A culture and politics of reviewing has existed almost as long as book publication has been noted in the press, but scholarship in the area is limited, particularly with respect to twentieth-century popular media and books. In fact, in my work, the distinction between book reviewing of *Silent Spring* and general coverage of the debate was one of the first boundaries to dissolve.

In her classic essay "Reviewing," Virginia Woolf noted that book reviewing "came into existence with the newspaper."[29] She differentiated criticism from reviewing thus: "The critic dealt with the past and with principles; the reviewer took the measure of new books as they fell from the press."[30] In his attempt to distinguish reviewing from literary criticism, Granville Hicks coined the term "literary journalism."[31] Other journalists also grappled with the genre, sharing Adolph Ochs's view that books can be newsworthy. "A book-reviewer is partly a purveyor of news," Clifton Fadiman put it succinctly, and John Drewry offered a valuable description of reviewing's journalistic function: "The review is like the editorial in that it is interpretative and explanatory."[32]

Joan Shelley Rubin confronted the dual traditions directly: " 'The question of whether attention to new books should take the form of 'news' or 'criticism' . . . shaped American book reviewing since the nineteenth century." The "news approach" was, according to Rubin, in constant tension with other impulses, including pressure from publishers who were also advertisers. Eventually (following the British example of the quarterly or monthly reviewing periodical), reviewers "willingly became journalists but they regarded their calling as 'higher' than that of the ordinary newspaper"—or as Robert Kirsch defined the book beat, "superior cultural reporting."[33] In Rubin's discussion of "middlebrow" taste, she offered, in effect, a continuum between treatment of books as news and exercise of literary criticism, allowing for variations in degree but avoiding partition. The idea that either books *or* reviews might make news, let alone how they might do so, has otherwise been largely absent from academic literature, despite the almost daily appearance of comment on books.[34]

Meanwhile, in practical descriptions of the book industry, book reviews have customarily been included for their significance in the marketing effort without much scrutiny of the reviewing process and its significance beyond selling a book to periodical subscribers. Yet once again, practitioners' writings are often an important source of historical information on the politics and culture of reviewing, especially when written by industry insiders—often in the context of complaints about deterioration in the quality of reviewing.[35] Book review editors wield considerable power simply in the choice of books to review or ignore. Their choice of reviewers, furthermore, may follow politicized patterns that, for example, match J. Bowman's "ideologic twins," as when like-minded authors review each other's books, rendering reviews bland or worse.[36] Or, by the same token, editors may counterpose antagonists having "bellicose political agendas" for dramatic or even partisan effect.[37] Even without editorial intervention, reviewers who review each other may indulge in such political foibles as reciprocation for good will or retaliation for ill will.[38] A helpful overview of reviewing and reviewing culture is the chapter on book reviewers in Coser, Kadushin, and Powell's 1982 work, admittedly now rather outdated in this era of the book tour and talk show; but they describe many aspects of reviewing that were nonetheless still relevant to the *Silent Spring* era.[39] The chapter notes the significance of what is chosen for review, where the review appears, who the reviewer is, timing, and type of medium, with particular attention to the power of prestige media, the *New York Times* above all.

The enormous and perhaps disproportionate power of the *New York Times* among review organs is the subject of a number of discussions of book reviewing, including Ann Haugland's investigation of the cultural significance of the *New York Times Book Review* for American readership.[40] By contrast, it is difficult to find much discussion at all, even in the trade magazine *Publishers' Weekly*, of the role of hometown or regional newspaper reviewers, whose importance for book promotion was nonetheless well acknowledged at the time of *Silent Spring* by book agents, publishers' publicists, and the company's "travelers" purveying the season's offerings around the country.

Any such study of regional or local reviewing, including local broadcast reviewing, would necessarily trip over an economic function of reviewing rarely made explicit, even with respect to elite publications such as the *Times* or the *New Yorker*: that of *selling the medium and the readership to advertisers*. Quoting reviewer Harry Hansen, Rubin touched on this with respect to book publishers who wished to advertise but also with respect to advertisers of other, often luxury products: "[Newspaper publishers] found that the book section could be used as a quality argument with other advertisers."[41] That function puts a remarkable spin on the cultural effect of reviewers' choices; and R. S. Fogerty's description of the role of the review—"to create an audience for a good work"—raises a significant but complicated question of whose tastes actually prevail in the marketplace.[42]

In a similar vein, one additional function of the book review seems at first glance incidental, but placed in the larger context of the entire media system, it acquires considerable significance. From its earliest forms, book reviewing has often been used to stand in for the book itself. Nina Baym made the point glancingly, noting that early reviews carried long synopses precisely because some periodical readers would not have access to the book itself.[43] Drewry listed among several groups of book-review readers, "those who do not have the time to read books and who must therefore rely on reviews for their information."[44] As my study has shown, the ability of reviews not only to represent the book but also to micro-mirror an entire debate emerged as a pivotal element in the public life of *Silent Spring*.

How readers respond to books is the focus of readership and reception theory, an increasingly well represented field in book history studies (and one related to audience reception study in mass communication). In his discussion of the "order of the book," Roger Chartier discussed "communities of readers"; and others have addressed similar organizing principles regarding readership, using groupings of readers with shared perspectives causing shared interpretation of meaning and shared relation to the social order.[45] The significance and function of reading in American culture has been explored rather more thoroughly with respect to pre-twentieth-century readers than to later audiences; and the responding reader appears in works such as David Hall's book on colonial religion and reading practices, Michael Warner's discussion of publication and the public sphere in the eighteenth century, and Richard Brodhead's study relating nineteenth century reading to print and publication.[46] Cathy Davidson, in particular, has written much on the subject of reading, but like many who follow her, she is most concerned with the role of fiction.[47]

The emphasis on fiction and "serious" literature has been mitigated by the presence of class as a research issue, from Laurence Levine's "highbrow/ lowbrow" distinction and Joan Rubin's significant study of books and "middlebrow culture" to the establishment of popular culture as a valid subject of investigation.[48] In their considerations of the Book-of-the-Month Club, Joan Rubin and Janice Radway each explored the concept of "serious" reading for the non-highbrow reader. Radway then moved further into the realm of mass-audience reading of romance novels, relevant to my work because her orientation was to discover, in effect, how the romance readers "used" the books.[49]

The role of nonfiction, "journalistic" books, has been rather neglected, in part because they are largely a recent phenomenon and in part because they are not usually within the purview of literary studies. Studies of books' role in social change do exist, though often surrounding fiction that challenges mores and morality. An exceptional example is Darnton's *Forbidden Best-Sellers of Pre-Revolutionary France*, in which he explores the trade in illegal best-sellers and its relation to the ideologies and attitudes that led to the French Revolution. He describes the "culture of dissent" that was evident in this trade and notes particularly the "*libelles*," underground books exposing the misdeeds of those in power, which he calls "a kind of journalism disguised as contemporary history and biography."[50] Moreover, the third section of his book asks, "Do Books Cause Revolutions?" For an answer, Darnton prescribes a melding of approaches based on discourse theory—referring to public processing of ideas—with those based on anthropology's diffusion theory, following trails of dispersion of books and/ or their ideas; his prescription informed certain aspects of my work. Darton notes that ideas are not unitary and that tracing their diffusion as if they were is untenable (an argument parallel to criticisms of the early "silver bullet" model in communication theory). He nonetheless points out that books *are* unitary and can provide the basis for both quantitative study and descriptive history.

Otherwise, the idea that books may be influential in history—or at least may influence participants in history—is also implicit in the time-honored practice of creating a list of books-that-mattered, most-important-books, great best-sellers, and the like. The *New York Public Library's Books of the Century* is a prime example, and many like it have included *Silent Spring* in their lists, particularly as the twentieth century drew to a close.[51] The category of nonfiction did not appear on such lists until 1917 and has tended to be dominated by memoirs, biography, and how-to books. Frank Luther Mott's *Golden Multitudes* describes the rise and nature of the best-seller phenomenon up to the late 1940s, but fiction again predominates in his discussion with only passing nods to religious works and other nonfiction.[52]

Studying books as news makers or as instigators of media-borne public debate nonetheless still requires accounting for books as part of a media system, which print culture scholarship has tended to sidestep despite its umbrella interest in all printed communication. The usual starting point is to see books as one among several media competing for the mass audience's time and attention, as Peter Mann discusses briefly in his social study of British publishing history. He writes: "In the printed world alone both newspapers and magazines are very serious competitors which cater extremely skillfully for the person who has either little time or little inclination for the volume of content which the book offers. Newspapers today, particularly the more serious ones, contain so many feature articles that the general reader may feel that he has sufficient information on, say, the problems of central Africa without going to a book-length analysis."[53] Again, the potential for positive or synergistic interaction between books and other mass media has not been not considered by Mann or other writers, with two exceptions, one being Darnton's complicated model of interactions among pre–French Revolution communication vectors (that is, pre–mass media), the other being Thomas Whiteside's analysis of books' function as "properties" taken up by magazines, film, and television through assignment of rights.[54]

Although most recent considerations of reader response assert that the reading act is not a passive one, most are more concerned with how the readers' personal or cultural settings affect reception of a textual message than with the *result* of that reading (beyond abstract interior activities such as self-fashioning). Such analysis thus stops short of considering that readers might actually take action in response to a book, although a number of scholars have looked at the specific action of writing personal letters to an author.[55]

There is also a small but growing body of literature on letters to the editor as a gauge of public opinion, notably work by David Paul Nord and Brian Thornton, although these most often concern attitudes toward journalism itself, rather than the issues that stimulated the writing response in the first place.[56] Two who have specifically addressed letter writing as a public gesture in the context of social activism are David Thelen, who wrote about letters written to Congress regarding the Iran-Contra controversy, and Karin Wahl-Jorgensen, whose writings on the letters-to-the-editor page as a forum for public debate have retroactive pertinence to the public conduct of the *Silent Spring* debate.[57] Although neither Thelen nor Wahl-Jorgensen dealt with the debate's point of origin (which in the case of the 1962–63 pesticide controversy was a book), their acknowledgment of letter writers' expectations had direct bearing on understanding what *Silent Spring's* letter-writing audience thought they were doing.

Asking questions about the point of origin of the debate brings us to rather hoary concepts from mass communication studies that nonetheless help to describe the interaction between the book *Silent Spring* and the media—as well as to concepts that begin to answer the question of whether books will die. Despite the ubiquity of books in chain stores, shopping malls, supermarkets, drugstores, discount houses, train stations, and airport newsstands (not to mention the number of books written by scholars in mass communication and journalism history), the field of mass communication studies has rather studiously avoided consideration of books as a mass medium. The general failure to include books in the scope of mass communication analysis has thereby left noteworthy gaps in the field in general and in media and journalism history in particular.

The concept of agenda setting is so well established in mass communication studies that it is almost considered passé except among those who have teased further concepts out of it. The media's ability to focus public attention was identified as early as 1922 by Walter Lippmann and was later discussed by Harold Lasswell in his 1948 work on communication's social functions, and by Paul Lazarsfeld and Robert Merton in their reconsideration of propaganda and the relationship of the media to social change.[58] In 1963 the idea was articulated more explicitly by Bernard Cohen: "The press is significantly more than a purveyor of information and opinion. It may not be successful much of the time in telling people what to think, but it is stunningly successful in telling its readers what to think *about*."[59] This effect was termed "agenda setting" and operationalized in a study of voter perceptions by Maxwell McCombs and Donald Shaw in 1972.[60]

Within the concept of agenda setting, several questions must be raised. First, what constitutes an agenda item—what becomes "grist" for the media mill? And if the media indeed set the public agenda, who or what proposes an issue to the media—what sets the process in motion? Finally who, what, or where is the "public"?—a question that takes the discussion out of agenda setting and draws it into the more generally sociological as well as the political-philosophic.[61]

Everett Rogers and James Dearing defined "agenda" as "a list of issues and events that are viewed at a point in time as ranked in a hierarchy of importance."[62] Defining the content of a public or media agenda as *both* issues and events presents the further challenge of deciding how to distinguish issues *from* events and yet relate them in a way that makes research and analysis viable. The distinction between issues and events reflects an important legacy of communication studies' journalistic origins insofar as the field has long recognized that events are more readily handled by reportorial journalism, while issues are difficult to report on and are more likely to spring to life in editorials and commentary. Some scholars have tried to define an issue as a series of events related by a broader idea;[63] but the difficulty is, as Lippmann put it, that "before a series of events become news they have usually to make themselves noticeable in some more or less overt act."[64] In 1981 Kurt Lang and Gladys Engel Lang sought to taxonomize the various conceptualizations of what an issue is for these purposes, acknowledging that the existence of controversy and policy implications may be key components.[65] For his study of coverage of 1960s issues, G. Ray Funkhouser distinguished between event-based issues and non-event-based issues. Non-event-based issues were those not inherently "reportable as news" but able to enter the news through "pseudo-events" or situations that "summarize non-newsworthy events in a newsworthy way." He called these situations "event summaries"[66]—a term readily applicable to a book that pulls together collected information and reported events in a thematically unified whole, as *Silent Spring* did. Furthermore, the dependence of the news media on Funkhouser's "event summaries" is particularly relevant to scientific news.

Science is often difficult to report and inaccessible to the mass audience, especially when intricate relationships, long-term processes, and technical methodologies are involved. Writings in environmental communication have tended to focus on the difficulty of communicating ecological issues in compelling ways, often resulting in a concentration on rhetoric or overlooking dynamics outside the field of environmental studies. In Anthony Downs's study of ecology as a public issue in the 1960s, he discussed, as others have, the need for a newsworthy issue to be dramatic and easily perceived.[67] As part of a remedy, study of agenda setting has been augmented by the concept of *agenda building*, which better reflects the evolution of public debate on an issue and also begins to account for the extra-media forces seeking to have an impact on the public agenda (notably, lobbying and public relations activities not directed at the media). Lang and Lang's study of agenda building describes the process of linking an issue to the political landscape and the rise of "spokesmen who can articulate demands" through commanding media attention. "The process is a continuous one, involving a number of feedback loops."[68]

The circularity of their model hints at a systems-theory approach to the process, by which public discussion is characterized as a system of systems, much as the media system is a system of systems within the broader sociopolitical systems. The tendency for media to draw on one another for content also evokes Marshall McLuhan's perception that one medium is the content of another.[69] Nowhere is that more true than with books that make news—as *Silent Spring* did—as they are taken up by reviewers and commentators, not to mention reporters. In such cases, the books constitute both newsworthy events and event summaries; and either way they become agenda items for media, public, and policymakers.

Lest analysis get lost in circularity, the second question remains: If the media set and/or build an agenda for the public, who or what sets the agenda for the media? The question has been raised in a number of studies of agenda setting, though few address it directly.[70] Lippmann compared issues not linked to events of guaranteed newsworthiness to unscored ballgames in which the press is umpire.[71] Obviously, circumstance itself can determine what the media will discuss: a terrorist attack will prompt discussion of terrorism. Yet even "reality" as a variable poses some problems. Funkhouser, for example, had difficulty operationalizing the concept, using statistics such as number of crimes per 100,000 inhabitants to indicate crime, number of troops and battle deaths to quantify the degree of Vietnam involvement, and so forth—an approach he himself questioned.[72] Nonetheless, events will always provide ready agenda items on the media's "budget," while the stimulus of *issue* coverage in the absence of a discrete event is often much less easily identified. When Bruce Westley asked "what makes it change?" ("it" referring to the agenda-item hierarchy), his answer included the actions of institutions and special interest groups, as well as "what's happening, even when it's being made to happen."[73]

In their very brief review of the literature on "who sets the media agenda,"[74] Rogers and Dearing dealt with answers that operate largely at the institutional level, such as governments and agencies. Although structural analysis may suggest why an issue is chosen for attention and how it is treated, it does not always explain initiation or source, particularly of controversies challenging the moneyed status quo. As Lucig Danielian and Stephen Reese put it, Rogers and Dearing "left no room for influences from sources who seek to purposefully influence media agendas. Therefore we propose adding sources as a factor in media agenda setting." In doing so, they identified the further question of "which sources get to set the media's agenda." It troubled them that "for the cocaine story, it appeared to be mostly national leaders," for if national leaders are the only voices heard, the news remains solely event-driven and state-determined, and there is no public debate.[75]

The idea that agenda items might be manipulated and fed to the media by agents with special purposes gains considerable weight in the study of professionalized agenda building—namely, public relations—and in light of what Oscar Gandy termed "information subsidies." Also concerned with who sets the media's agenda, Gandy offered the proposition that distribution and circulation of information is affected by the "cost" of that information—in other words, the amount of time, effort, or money required to obtain it: "Some information, like advertising and other promotional messages, is over-produced, and is provided free to its consumers."[76] To the extent that the information provided is free or made available at much less cost to journalists than it would have cost them to gather such "news" on their own, the sources have created an "information subsidy" to the press.[77]

The applicability of the concept of information subsidy to the *Silent Spring* episode is inviting. However, a book is more than and different from an exceptionally long press release, and its role in the agenda-setting process is as complex as the process itself. In Philip Meyer's call for greater responsibility on the part of journalistic book authors for accuracy and reasonableness, he acknowledged that the "greater flexibility" of book publishers does enable book publishers "to perform an air-clearing and agenda-setting function for topics that newspapers will not touch without a certain amount of nudging," making "an important contribution to pluralism, creating multiple paths to agenda setting and bypassing the rigidity of newspapers." The "rigidity" of newspapers, for Meyer, derives from an odd linkage between attractiveness to advertisers and a reputation for credibility; and it was Meyer who observed that because of changes in publishing technology, "the book author has become the modern equivalent of the lonely pamphleteer."[78] Jürgen Habermas, among many others, might well argue that credibility and attractiveness to advertisers are antithetical; but paradoxically (and regardless of production technology), the very fact that books are—or were in the 1960s—the one medium not dependent on advertising revenue does support the image of the book author as the lone pamphleteer.

For an author, the point of writing a book like *Silent Spring* is to disseminate information. A few ideas from within the study of information flow are therefore applicable here. Traditional mass communication studies of information flow, such as Rogers's study of diffusion of innovation and Katz and Lazarsfeld's *Personal Influence*, have been concerned primarily with the relationship between the media and the receiving individual rather than with information flow *among* media or from individual *to* media.[79] More often, studies of the process have been more concerned with discerning trajectories of influence than with tracking information and any alterations in it across media. Quite often, the studies suffer from two basic liabilities of quantification: complex or descriptive information is lost, and chronology is presumed to mean causality in the classic "post hoc" fallacy. The "life course" of an issue, for example, is often examined through quantification of its "appearances" in public places—such as number of column inches or number of stories—and is plotted over time to arrive at correlated changes in agendas.[80] Funkhouser's study of the rises and falls of public and/or media attention to certain 1960s issues, including the environment, presented him with the problem of accounting for peaks of interest not directly associated with newsworthy events, which foreshadowed the problematic differences between events and issues and the entire idea of agenda building: "A case can be made for explaining many of the peaks in coverage of non-event based issues as being related to 'manufactured news' rather than 'spontaneous news.' "[81] The appearance of Carson's book and response to it could be said in some ways to fall into the category of "manufactured" news regarding the environment, though he did not extend his discussion to that possibility. Furthermore, analysts of the life cycle of a public issue tend to be more interested in its demise than in its birth.

Frequently, analyses are directed at correlations between a summarizing version of media agendas (for example, all news stories in print media or total frequency across news programs of three networks) and non-media agendas—most commonly the "public" agenda and the "policy" agenda.[82] Interactions *within* the media system are particularly pertinent to the *Silent Spring* episode, and some have studied what Danielian and Reese termed "intermedia" interactions.[83] But once again we must note that media interaction is more than a simple, unidirectional flow of influence or information. In his study of print and television in West Germany, Klaus Schoenbach suggested a possible potentiation of "salience" when both media are involved: what may have been salient in one or the other medium became more so when it appeared in both.[84] Certainly, the *Silent Spring* experience demonstrated this phenomenon, but at the same time, the concept comes perilously close to saying that something makes news because it makes news.

Yet Schoenbach's observations hinge in part on how much information any given medium can carry; and the concept of information or channel capacity is an intriguing one here. Marc Benton and P. Jean Frazier identified three levels of "information holding" that were seen to affect differently the direction of influence and the longevity of an issue on the public agenda.[85] Considering that books commonly occupy the far end of a continuum of information capacity, the special ability to carry a lot of information becomes quite significant in the study of books' role in a public controversy. The suggestion of both studies was that television—having limited information capacity—is important in the agenda-setting process when issues are event-driven, while print media—of greater capacity—have a more important role in non-event-driven (that is, issue) agenda setting and are likely to sustain a longer "life" for the issue. There was some evidence of that process in the case of *Silent Spring,* but in an important way the model was turned somewhat inside out insofar as national television was effectively the last medium to join the debate.

Before moving to the broad question of where, exactly, the debate took place, one last area of mass communication theory must be mentioned with respect to the "death" of books. The field of study commonly termed "uses and gratifications" addresses why people "use" a given medium, considering not only what a book is but also what it does; and analysis has ranged from ergonomics to philosophy. Most of the writing in this field concentrates on the audience, most frequently the individual audience member's needs and uses of one medium or perhaps two—with some acknowledging the need to relate the individual's "use" of a medium to social context.[86] Some literary studies have addressed how readers "use" books, and Radway's studies of romance reading stand out among them.[87] Mass communication studies, however, do not commonly compare audience choices among media except regarding news consumption. Further, with a single exception from the 1970s, those studies are unlikely to examine the particular traits of a given medium, toward understanding *why* an individual or a community might choose one medium over another— for example, film for entertainment versus radio or television for breaking news. That lone exception, a 1973 study of an Israeli community by Elihu Katz, Michael Gurevitch, and Hadassah Haas, was one of the very rare studies to include books among mass media studied.[88]

The medium of books is otherwise almost universally excluded from analysis, in part because of the preference for studying news consumption, in part because of methodological habit. Studies tend to select only two or at most three media at a time, usually comparing periodical print media with broadcast media, thereby missing not only choices from among the full array of the media system but also interactions between them, as with cross-media advertising or, notably, book reviewing. We are, after all, consumers of multiple media, reading print media, watching broadcast media, and listening to audio media all in the same life course. Yet the recent, feverish study of audience effects and interactivity typically sidesteps the compounding effects and interactions of all these concurrent mediations of information and experience.

A compelling and often unposed question about any medium in this light is therefore not what it *is* but what it *does* for the audience, as well as for the other media in the media system. Yet much of the "end of the book" discussion devolves to the question "what is a book?" Is it a bound volume of paper covered with words and/or pictures? Is it a chunk of text larger than a short story but not so large that it cannot be read in, say, a few days? Is it a concept, impression, or story by a single author of a given complexity and breadth, or is it a collectivity of thought by an infinitely expandable company of authors, editors, and readers? Defining the entity that can be called a book has occupied several writers in literary and critical studies,[89] yet most analyses are descriptive or philosophical, with functional definitions left inchoate in reading and reception theory. Meanwhile, the study of mass communication—in an excellent position to offer more functionalist definitions—rarely mires itself in such definitional pursuits. Having explored what the book *Silent Spring* "did" for producers, distributors, detractors, and audience, we can begin to see the possible richness of a more functional approach to media study in general and to book history as well. Books, as I discussed earlier, are "used" for many purposes; thus, for them to "die" as a medium, those purposes would have to disappear.

Lastly, returning to the mission of a work of "literary nonfiction designed to raise public consciousness," let us remember that the audience for whom Carson wrote her book was, indeed, the public—not merely a universe of individual readers but a franchised polity whom she hoped to influence. The version of systems theory adopted within mass communication theory says that communication's function is to provide the "feedback" that supports and maintains the sociopolitical system within which the media system operates. In less theoretical terms, that means that communication media can work either to maintain the status quo or to effect social change—whichever is needed for the survival of the larger socioeconomic system of systems.

Historically, at least in America, the media system has indeed been a major venue for the play of dynamics between stasis and change. In the real-world context of actual feedback and real agendas, analysis must recognize the realm of the public: the readers, viewers, and listeners who are also consumers and voters, in the context of their implied relationship to the political system. Studies of agenda setting that concentrate on correlations between media agendas and policy are based on a methodological partition of the media from the public, in a paradoxical relationship to each other with the subtext that the media should be, but may not be, a legitimate voice of the people. Exploring the public career of *Silent Spring* thus quickly demands asking about the usefulness of that partition, especially regarding public discourse in the form of active debate. When brought to the "real" and mass media worlds, the fluidity of the concept of "public" becomes yet more problematic as it glosses over the actual perceptions and multiple roles of the audience. That fluidity was well demonstrated when the reader-writers, occupying an amorphous zone between individual citizenship and public collectivity, took up pen or telephone to become themselves part of the media response to *Silent Spring*.

Jürgen Habermas's simplest definition of "public sphere" has been expressed as "a domain of our social life in which such a thing as public opinion can be formed."[90] The historical existence, duration, and nature of a true public sphere in the various eras of both premodern and modern society have nonetheless been matters of vigorous debate (even between Habermas and himself). A defining trait of Habermas's public sphere is, however, the opportunity for "rational-critical" political discourse, on the basis of which policy and state behavior are expected to evolve in the public interest. His model would seem to embrace the Miltonian "marketplace of ideas" undergirding the American democratic paradigm. American democratic ideology has, after all, insisted on the presence of political elasticity; and reverence for the First Amendment has been rooted in political dogma protecting public dissent and argument for the very purpose of preserving the democratic system. The existence of a public controversy is prima facie evidence that the system is working.

Habermas and others have, however, detected distortion of a true public sphere by private involvement. To Habermas, the current era in which the privately owned mass media are public participants represents a deterioration in "rational-critical" discourse, insofar as the democratized public sphere has degenerated into a locus of "uncritical mass consumer culture."[91] Because money and power are "non-discursive modes" of sociopolitical organization, it would follow that to the extent that the mass media represent profit-motivated private interests, the quality of public discourse degenerates.[92] Writing at the same time as Habermas in his early work, Jason Epstein, vice president of Random House, warned specifically about such incursions on book publishing's cultural and intellectual independence, especially in terms of constraints on risk taking.[93]

Traditionally, the most worrisome constraints on free exchange have been thought to be those imposed by government, with the press functioning as primary watchdog and outlet for the public voice. In that ideal, journalists are essential members of the public, not outside the public sphere but in fact manifestation of it. Raising some of the same questions as Habermas, Epstein, and others, Michael Schudson has discussed the effect of private ownership on American journalism—specifically the opportunity for controversy. He wrote: "While the commercial press is not without its virtues, actively engaging the public in political debate is not one of them."[94] Schudson, however, connected the function of news to the existence of a public forum, defining news as "public knowledge"; and he pointed out that bringing information into the public sphere does more than perform a simple act of communication. "When the media offer the public an item of news, they confer upon it public legitimacy."[95] That legitimation is necessary for the issue to be placed on the agenda of public debate and, further, to command the attention of those in power. In fact, "so long as the information is publicly available, political actors have to behave as if the public were paying attention."[96] Carson certainly counted on the perception if not the reality that the public was paying attention; and Schudson's very applicable perspective clearly influenced this analysis of *Silent Spring*'s public career.

Calhoun also challenged Habermas's degenerative model for its failure to consider whether social change might not only still be possible but perhaps even inherent in the system as it responds to the force and value of social flux, regardless of public or private ownership; and he cited the 1960s era as illustration.[97] The Miltonian "marketplace" imagery is indeed often declared to accommodate the possibility that even profit-driven, mediated communication offers choice to the consuming audience/polity— with the faith that a reasonable citizenry will perceive the virtuous and the valid, rejecting what will not sustain the system.

In Habermas's ideal, of course, private ownership of the part of the public sphere that flows through the mass media fatally taints "rational-critical" public discourse. Habermas's original formulation was written, notably, at the same time *Silent Spring* was in production. For Habermas in 1961, the rise of the field of public relations was emblematic of the disappearance of a true public sphere.[98] Strictly speaking, his interpretation of the fact that a controversy did arise around pesticides might be that the debate was profitable for the media participants and that therefore the debate was nothing more than a media commodity itself. As we review the motives and missions of everyone involved, the validity or irrelevance of such an interpretation is worth pondering.

Habermas did eventually acknowledge the possibility of the "social integrative power of communicative action" in a world that includes an influential, privately owned mass media. He was still concerned, however, about the "two crosscutting processes" of "communicative generation of legitimate power": the potential forces for social change versus "the manipulative deployment of media power to procure mass loyalty, consumer demand, and 'compliance' with systemic imperatives" in order to sustain the status quo. He wrote: "This is the question of whether, and to what extent, a public sphere dominated by mass media provides a realistic chance for the members of civil society, in their competition with the political and economic invaders' media power, to bring about changes in the spectrum of values, topics, and reasons channeled by external influences, to open it up in an innovative way, and to screen it critically."[99]

For Habermas, if the media can accomplish change, they will do so through the "power of public discourses that uncover topics of relevance to all of society, interpret values, contribute to the resolution of problems, generate good reasons, and debunk bad ones. . . . Discourses do not govern. They generate a communicative power that cannot take the place of administration but can only influence it." In his view, the impact of such "discourses" is achieved "in a siegelike manner."[100] Although the scope of this analysis of the *Silent Spring* debate ends roughly where the political results of the debate begin, my hope is that it nonetheless provides a vivid illustration of just such a "siege."

NOTES

Abbreviations Used in the Notes

Archives

RCP–BY   Rachel Carson Papers, Yale Collection in American Literature, Beinecke Library, Yale University, New Haven, Conn. Used by kind permission of Frances Collin; excerpts from unpublished and oral Rachel Carson material © 2000 by Roger Christie. Reprinted by permission of Frances Collin, Trustee. All rights reserved. Cited as RCP-BY followed by box number, e.g., RCP-BY 87.

HM–HL   Shelf mark *bMS Am 2105(42)*, Houghton Mifflin Archive. Used by kind permission of Houghton Mifflin Company and Houghton Library, Harvard University, Cambridge, Mass. Cited as HM-HL followed by file and box number, e.g., HM-HL AM10/3.

NYA–NYPL   *The New Yorker* Archives, New York Public Library. Used by kind permission, *The New Yorker* Records, Manuscripts and Archives Division, New York Public Library, Astor, Lenox and Tilden Foundations. *The New Yorker* material used courtesy of *The New Yorker*/Condé Nast Publications Inc. www.newyorker.com. Cited as NYA-NYPL followed by box number, e.g., NYA-NYPL 1292.

NARA   Records of the Office of the Secretary, Department of Agriculture, National Archives and Records Administration,

College Park, Md. Included is material from General Correspondence 1906–1975, Insecticides, Record Group 16, boxes 3782 and 3959, and General Correspondence to the Secretary of Agriculture, Publications, Record Group 310, box 651. Cited as NARA followed by record group and box numbers e.g., NARA RG 16, 3782 or 3959, and NARA RG 310, 651.

PR/DNI–HM&L   Public Relations files, E. I. DuPont Nemours, Inc., Hagley Museum and Library, Wilmington, Del. With gratitude for their use.

Correspondence

RC   Rachel Carson
MR   Marie Rodell, literary agent for Rachel Carson
PB   Paul Brooks, editor in chief, Houghton Mifflin
AF   Anne Ford, publicity, Houghton Mifflin
LT   Lovell Thompson, vice president for trade books, Houghton Mifflin
Note: writers of published letters to the media are identified by name; writers of unpublished letters are identified by initials and city of origin.

Chapter 1

1. Philip Meyer, "Accountability When Books Make News," *Media Studies Journal* (Summer 1992), 138.
2. The *New York Times* suffered a strike in the winter of 1962–63, and no *Times* bestseller list was available for inclusion in national best-seller calculations. Whether *Silent Spring's* tenure at the top of those lists would have been longer had a *Times* list been included is impossible to determine, but *Silent Spring's* strongest draw at the time was in the Northeast.
3. Based on *Publishers' Weekly* weekly lists for 1962 and 1963.
4. Rachel Carson, *Silent Spring* (Boston: Houghton Mifflin, 1962), 1–3.
5. *Silent Spring*, 12–13.
6. *Silent Spring*, 13
7. *Silent Spring*, 42.
8. *Silent Spring*, 64.
9. *Silent Spring*, 103.
10. *Silent Spring*, 152.
11. *Silent Spring*, 181.
12. *Silent Spring*, 259.
13. *Silent Spring*, 13.
14. *Silent Spring*, xi.
15. *Silent Spring*, 12.
16. See the particularly good account in "Poisons, Pests, and People," a film by the National Film Board of Canada, producer David Bairslow, 1960.

17. Newton Minnow, 9 May 1961 speech to the National Association of Broadcasters Annual Convention; *Vital Speeches* 27 (15 June 1961): 534.

18. The number of Sunday editions rose from 545 in 1952 to 558 in 1962. Data from *The Mass Media: Aspen Institute Guide to Communication Industry Trends* (New York: Praeger, 1978).

19. Paul Boyer, *By the Bomb's Early Light: American Thought and Culture at the Dawn of the Atomic Age* (New York: Pantheon, 1985), 32.

20. Boyer, *Bomb's Early Light*, 62.

21. Adam Rome has a particularly good discussion of the rise of environmentalism in the sixties and relates Carson's role to the role of women in the movement: " 'Give Earth a Chance': The Environmental Movement and the Sixties," *Journal of American History* 90 (September 2003): 525–54. See also Edmund Russell, *War and Nature: Fighting Humans and Insects with Chemicals from World War I to Silent Spring* (Cambridge, U.K., and New York: Cambridge University Press, 2001).

22. See, among others, Christopher J. Bosso, *Pesticides and Politics* (Pittsburgh: University of Pittsburgh Press, 1987); Frank Graham, *Since Silent Spring* (Boston: Houghton Mifflin, 1971); and James Whorton, *Before Silent Spring: Pesticides and Public Health in Pre-DDT America* (Princeton, N.J.: Princeton University Press, 1974).

23. John K. Terres, "Dynamite in DDT," *New Republic*, 25 March 1946, 415.

24. Murphy v. Butler, 270 F. 2nd 419, *cert. denied*, 362 U.S. 929 (1960).

25. William O. Douglas, "Dissent in Favor of Man," *Saturday Review*, 5 May 1960, 59-60. It was also reprinted in *U.S. News & World Report (USN&WR)*, 23 November 1959, 143.

26. "Controlling the Pesticides," editorial, *New York Times*, 31 July 1961, 18.

27. "Pesticides Are Good Friends, But Can Be Dangerous Enemies if Used by Zealots," editorial, *Saturday Evening Post*, 2 September 1961, 8.

28. Valerie Gunter and Craig Harris, "Noisy Winter: The DDT Controversy in the Years before *Silent Spring*," *Rural Sociology* 63 (June 1998): 197–98.

29. Habermas's simplest definition of "public sphere" has been expressed as "a domain of our social life in which such a thing as public opinion can be formed"; "The Public Sphere," in Chandra Mukerji and Michael Schudson, eds., *Rethinking Popular Culture: Contemporary Perspectives in Cultural Studies* (Berkeley: University of California Press, 1991), 398. See further discussion of Habermas and this concept in the appendix.

30. Kenneth Smith, "Rachel Carson's Curse," editorial, *Washington Times*, 10 February 2000.

Chapter 2

1. Background on Carson was collected from numerous sources, but Linda Lear's biography, *Rachel Carson: Witness for Nature* (New York: Henry Holt, 1997), provides an exceptionally rich, well researched, and authoritative portrait of Carson. Also, Paul Brooks provided considerable insight as to her work habits and professional style in *The House of Life: Rachel Carson at Work* (Boston: Houghton Mifflin, 1972). See the appendix for other sources.

2. RC to MR, [no month] 1955, HM-HL AM10/2.

3. RC to MR, 2 February 1958, RCP-BY 105.

4. RC to E. B. White, 3 February 1958, RCP-BY 44.

5. RC to MR, 3 April 1958, HM-HL AM10/2.

6. Carson expected his contribution to be limited to coverage of the Long Island trial and research in government records: "Mr. Diamond agrees to cooperate in every way possible in making available to me material that, for one reason or another, is less easily

accessible to me than to him (e.g. excerpts from the trial testimony, or certain other documents not available here)"; RC to MR, 9 April 1958, HM-HL AM 10/2. Diamond was displeased by the circumstances surrounding the termination of their association. He received a lump-sum payment and demanded the return of all research materials he had forwarded to her; letter from Edwin Diamond to PB, quoted in a letter written to the editor of the *Saturday Evening Post* (not published), 4 October 1963, HM-HL 16.

7. President's Science Advisory Committee, "The Uses of Pesticides," Government Printing Office, 15 May 1963; hereafter PSAC Report.

8. RC, "Help Your Child to Wonder," *Woman's Home Companion*, July 1956, 25–27.

9. RC to MR, 14 April 1958, HM-HL AM10/3.

10. Speech to American Association of University Women, 8 December 1959; RCP-BY 99.

11. Lear, *Carson*, 184.

12. Brooks, *House*, 256.

13. "I am giving details to special friends like you—not to others, but I suppose it's a futile effort to keep one's private affairs private. Somehow I have no wish to read of my ailments in literary gossip columns. Too much comfort to the chemical companies!" RC to Marjorie Spock, 12 April 1960, RCP-BY 44.

14. The absence of the child's father was often explained using the phrase "through early loss." According to Lear, the child's mother had had an affair with a married man, who never had any relationship with the child; Lear, *Carson*, 214, 236–3.

15. *Always, Rachel: The Letters of Rachel Carson and Dorothy Freeman, 1952–1964*, ed. Martha Freeman (Boston: Beacon, 1994).

16. In an on-line interview, Lear has addressed this issue, apparently the first to do so publicly. Accessed at http://www.annonline.com/interviews/980518.

17. In one letter to Dorothy Freeman she expressed just such fear, calling her attention to a passage in a John Barkham column in the *Saturday Review* (in which he discussed rumors that the journalist Dorothy Thompson might have had a homosexual liaison). She wrote that it "really frightened me—I don't want to put it in writing but I'll just say that the same implication could be implied about our correspondence" and urged her friend to destroy a particular set of letters. Freeman, *Always*, 529.

18. Lear, *Carson*, 152.

19. When Brooks had to revise his estimate of initial print run up to 60,000 copies, Rodell told Carson she thought the number still fell far short of probable demand, as it indeed did; MR to RC, 1 August 1962, RCP-BY 105.

20. RC, *Silent Spring*, 3.

21. RC to E. B. White, 3 February 1958, RCP-BY 44.

22. Freeman, *Always*, 259.

23. RC to Paul Brooks, quoted in Brooks, *House*, 256.

24. RC, speech to Women's National Book Association, 15 February 1963, reported in Brooks, *House*, 228.

25. RC, *The Sea Around Us* (New York: Oxford University Press, 1951), 1961 edition, xiii.

26. RC, *Silent Spring*, dedication.

27. RC, "What's the Reason Why," *New York Times Book Review*, 2 December 1962, 3.

28. Senate Committee on Government Operations, Subcommittee on Reorganization and International Organizations, *Interagency Coordination in Environmental Hazards (Pesticides)*, 88th Cong., 1st sess., 4 June 1963, Washington, D.C., Government Printing Office, 206.

29. RC, speech to the National Parks Association, 2 October 1962, RCP-BY 101.

30. RC, "She Started It All—Here's Her Reaction," *New York Herald Tribune*, 19 May 1963, 24.

31. RC, *Silent Spring*, 152.
32. RC, interviewed on *CBS Reports*, "The Verdict on Rachel Carson's Silent Spring," broadcast 15 May 1963, produced and written by Jay L. McMullen, © Columbia Broadcasting System, 1963.
33. Lear, *Carson*, 399.
34. Lear, *Carson*, 375.
35. RC, speech, Garden Club of America, 8 January 1963, RCP-BY 101.
36. RC, speech, National Parks Association, 2 October 1962, RCP-BY 101.
37. "The Gentle Storm Center," *Life*, 12 October 1962, 105.
38. RC, "What's the Reason Why," 3.
39. RC to Lois Crisler, 7 May 1959, RCP-BY 102.
40. RC to PB, 6 April 1962, RCP-BY 87.
41. Frank Graham Jr., *Since Silent Spring* (Boston: Houghton Mifflin, 1970), 195.
42. RC, letter to the editor, *Washington Post*, 22 April 1953, A14.
43. Senate Committee on Commerce, *Pesticide Research and Controls*, 6 June 1963, 88th Cong., 1st session, on S 1250 and S 1251, Washington, D.C., Government Printing Office, 1965, 18.
44. RC, *Silent Spring*, 183.
45. RC, "She Started It All," 24. She continues, "This is pointed out by the [PSAC] panel, which now brands the government's mass eradication programs directed against the gypsy moth, fire ant, Japanese beetle, and white-fringed beetle as 'failures,' based on unrealistic concepts."
46. Carson wrote to Robert Strother, "roving editor" for *Reader's Digest*, "Do try to see the USDA's thoroughly dishonest film: Fire Ant, Friend or Foe? [She misquoted the title.] The Department has an enormous propaganda engine, and it is working full steam to protect a great array of vested interests." RC to Robert Strother, 9 July 1959, RCP-BY 42.
47. RC, *Silent Spring*, 162.
48. RC, *Silent Spring*, 174. She is referring unmistakably to Vance Packard's recent bestseller, *The Hidden Persuaders*, which revealed and described the use of subliminal advertising.
49. See especially Craig Waddell, ed., *And No Birds Sing: Rhetorical Analyses of Rachel Carson's* Silent Spring (Carbondale: Southern Illinois University Press, 2000).
50. Freeman, *Always*, 386–87.
51. Freeman, *Always*, 291.
52. *Silent Spring*, 37.
53. Freeman, *Always*, 290.
54. MR to RC, 2 December 1960, RCP-BY 387.
55. Interview with John Barkham, *Saturday Review* Syndicate, quoted in Brooks, *House*, 233. "I began to ask around for the information she wanted and the more I learned about the use of pesticides the more appalled I became. I realized that here was the material for a book." In fact, Huckins had written two letters, the first published in the Boston *Herald* and the second written personally to Carson with the *Herald* letter enclosed asking for information and contacts.
56. RC to S. C., Hamden, Conn., 28 March 1963, RCP-BY 88.
57. RC, speech, National Parks Association, 2 October 1962, RCP-BY 101 (original emphasis).
58. RC, speech, National Parks Association, 2 October 1962, RCP-BY 101.
59. Freeman, *Always*, 407.
60. Freeman, *Always*, 411.
61. Lear, *Carson*, 421.
62. MR to AF, Houghton Mifflin, 10 January 1963, RCP-BY 105.

63. MR to RC, 14 August 1962, RCP-BY 105.

64. "I must confess I am happy to say I am not yet free from CBS shackles"; RC to AF, 5 April 1963, RCP-BY 89.

65. Freeman, *Always*, 445.

66. Freeman, *Always*, 451. Lear reports that her condition was obvious and saddening to Sevareid and producer Jay McMullen: "McMullen remembers that as they were driving away, Sevareid agreed that they should get it on air as soon as possible. 'Jay,' he remarked, 'you've got a dead leading lady' "; Lear, *Carson*, 425.

67. RC to Clarence Cottam, 18 November 1958, RCP-BY 42.

68. RC to Clarence Cottam, 8 January 1959, RCP-BY 42.

69. RC, "Vanishing Americans," *Washington Post*, 10 April 1959, A12.

70. MR to PB, 24 June 1959, RCP-BY 87.

71. RC to Charles Callison, Audubon Society, 15 July 1961, RCP-BY 42.

72. RC to PB, 13 March 1963, HM-HL AM10/3.

73. Senate Committee on Commerce, *Pesticide Research and Controls*.

74. MR to RC, 2 December 1960, RCP-BY 105.

75. See H. Patricia Hynes, *The Recurrent Silent Spring* (New York: Pergamon, 1989), for a feminist perspective. She noted, for example, the clearly misogynist aspects of some criticism of *Silent Spring*.

76. See Lear, *Carson*, 429. Paul Brooks also referred to the incident without attributing it to Benson.

77. RC, speech, National Parks Association, 2 October 1962, RCP-BY 101.

78. MR to AF, 25 February 1963, RCP-BY 88. A Freedom of Information Act request yielded only one document concerning Carson from the Federal Bureau of Investigation, and it did suggest possible suspicion of Communist Party connections. The two-page dossier on Carson, dated 14 December 1962, had approximately 95 percent of the text blacked out. Codes explaining the censorship appeared in the margin, referencing "personal privacy" and "national security," along with the word "Russia." The dossier seemed to be noting at least three events: advice called to the FBI from the Immigration and Naturalization Service on 30 August 1962; further discussion (with the word "Russia" in the margin) on 27 November; and a telephone call to Carson's home on 19 November, although all other details were obliterated.

79. LT to MR, 14 August 1962, RCP-BY 88.

80. RC, speech, Women's National Press Club, 5 December 1962, RCP-BY 101.

81. MR to PB, 16 February 1962, and PB to MR, 9 April 1962, RCP-BY 87.

82. MR to AF, 20 April 1962, RCP-BY 87.

83. RC to Clarence Cottam, 17 January 1961, RCP-BY 42.

84. RC, speech, Women's National Press Club, 5 December 1962, RCP-BY 101.

85. RC to AF, 20 February 1963, RCP-BY 88.

86. Senate Committee, *Interagency Coordination in Environmental Hazards (Pesticides)*, 220.

87. The exception was Dr. Cynthia Westcott, academic entomologist and chair of the Garden Enemies Committee within the Council of Federated Garden Clubs.

88. RC, speech, Women's National Press Club, 5 December 1962, RCP-BY 101.

89. Speech, Garden Club of America, Vassar, Poughkeepsie, N.Y., 2 May 1963, HM-HL AM10/3.

90. MR to C.W.B., Vassar College, 2 May 1963, HM-HL AM10/3.

91. MR to RC, 31 August 1962, RCP-BY 88.

92. MR to PB, 16 October 1962, RCP-BY 105.

93. RC to PB and William Shawn, 14 February 1959, RCP-BY 87.

94. RC to William Shawn, 8 May 1963, RCP-BY 42.

95. MR to PB, 5 September 1960, HM-HL AM10/3.

96. RC, interview with Dr. Gabriel Fielding for BBC's *World of Books*, transcription from telediphone recording, RCP-BY 74.

97. RC to PB, 3 June 1959, RCP-BY 87.

98. Lear, *Carson*, 130.

99. Freeman, *Always*, 252 and 257.

100. Freeman, *Always*, 177.

101. Brooks, *House*, 133; Lear, *Carson*, 190, 238–40.

102. Correspondence between MR and Dr. L. A. Elvehiem, University of Wisconsin, 18 January 1963; H. B. McCarty to University of Wisconsin president Fred Harvey Harrington, 16 January 1963, Harrington to MR, 7 February 1963; MR to Harrington, 16 April 1963; RCP-BY 105.

103. Lear, *Carson*, 321.

104. RC to MR, 18 April 1956, RCP-BY 104.

Chapter 3

1. For histories of the *New Yorker*, see Ben Yagoda, *About Town: The New Yorker and the World It Made* (New York: Scribner, 2000); Gigi Mahon, *The Last Days of The New Yorker* (New York: McGraw-Hill Publishing, 1988); Brendan Gill, *Here at The New Yorker* (New York: Random House, 1975); Jane Grant, *Ross, The New Yorker, and Me* (New York: Reynal, 1968); Thomas Kunkel, *Genius in Disguise: Harold Ross of The New Yorker* (New York: Random House, 1995); and Ved Mehta, *Remembering Mr. Shawn's New Yorker: The Invisible Art of Editing* (New York: Overlook Press, 1998). See also discussions passim in John Tebbel and Mary Ellen Zuckerman, *The Magazine in America 1741–1990* (New York: Oxford, 1991); George H. Douglas, *The Smart Magazines* (New York: Archon Books, 1991); Theodore Peterson, *Magazines in the Twentieth Century* (Urbana: University of Illinois Press, 1956); Roland E. Wolseley, *Understanding Magazines* (Ames: Iowa State University Press, 1969); and James Playsted Wood, *Magazines in the United States* (New York: Ronald Press, 1971). See also Ben Yagoda, "An American Icon: High Seriousness to 1990s Buzz," *Chronicle of Higher Education* (31 July 1998): B6; and Trysh Travis, "The *New Yorker*: From Heroic Journalism to the 'Gentlest of Magazines,' " in "Reading Matters: Book Men, 'Serious' Readers, and the Rise of Mass Culture, 1930–1970" (Ph.D diss., Yale University, 1998), 125–69.

2. This statement is quoted in almost every work on the history of the *New Yorker* but taken here from Tebbel and Zuckerman, *Magazine in America*, 219–20.

3. Quoted in Mahon, *Last Days*, 4.

4. The notable exceptions are Yagoda's *About Town*, Mehta's *Remembering Mr. Shawn's New Yorker*, and the memoir of his long-term lover, Lillian Ross, in *Here but Not Here: A Love Story* (New York: Random House, 1998).

5. Gill, *Here at The New Yorker*, 150.

6. Gill, *Here at The New Yorker*, 25.

7. For 1970, Wood reported that 77 percent of the total circulation came from areas outside metropolitan New York (Wood, *Magazines*, 273). Mahon claimed 80 percent for the same era (Mahon, *Last Days*, 45). According to a *Time* article in 1960, the magazine had "97 subscribers in Dubuque, including several old ladies" ("The Years Without Ross," *Time* 75 [16 May 1960], 73).

8. Douglas, *Smart Magazines*, 171.

9. Grant, *Ross, The New Yorker, and Me*, 11.

10. Tom Wolfe of the *New York Herald Tribune* (and writing for the budding competitor

New York magazine) called Shawn an "undertaker" (Tebbel and Zuckerman, *Magazine in America*, 314), who had "essentially 'embalmed' the magazine and drained it of all its literary juices" (Douglas, *Smart Magazines,* 172).

11. Quoted in Gill, *Here at The New Yorker,* 391–92.

12. Quoted in Mehta, *Remembering Mr. Shawn's New Yorker,* 127.

13. Shawn felt he should publish material that would "make people reach for it, it's good for them." Whitman Hobbes, former full-time marketing consultant to the *New Yorker,* personal communication, 30 April 1997. See also Yagoda's discussion in *About Town,* chap. 6.

14. Comment, *Holyoke (Mass.) Transcript-Telegram,* 5 January 1963, RCP-BY 91.

15. Hobbes, personal communication, 30 April 1997. Many others also have reported this as an essential trait of the *New Yorker;* e.g., see Mahon, *Last Days,* 3, 36; Wood, *Magazines,* 272.

16. Mahon, *Last Days,* 5.

17. *New York Times,* 13 March 1962, 44.

18. Whitman Hobbes, personal communication, 27 October 1999.

19. Hobbes, 27 October 1999.

20. Travis, "Reading Matters."

21. Sources for the history of Houghton Mifflin include its own corporate history, accessed on-line 1 November 1999 at: http://www.hmco.com/hmco/corporate/history/History.html. In addition John Tebbel's histories include substantial information on Houghton Mifflin: *The Great Change, 1940–1980,* vol. 4 of *A History of Book Publishing in the United States.* (New York: Bowker, 1981), and *Between Covers: The Rise and Transformation of Book Publishing in America* (New York: Oxford University Press, 1987). For earlier years, see Ellen B. Ballou's *The Building of the House: Houghton Mifflin's Formative Years* (Boston: Houghton Mifflin, 1970). Houghton Mifflin is also discussed in the works of Charles A. Madison, *Book Publishing in America* (New York: McGraw-Hill, 1966), and in the earlier discussion of book publishing by Hellmut Lehmann-Haupt, *The Book in America: A History of the Making and Selling of Books in the United States* (New York: Bowker, 1951). See also Madeline B. Stern, *Imprints on History: Book Publishers and American Frontiers* (Bloomington: Indiana University Press, 1956), 354; and "The Story of Houghton Mifflin Company," *Book Production Magazine* 80 (September 1964): 30–34. Paul Brooks's more personal memoir in *Two Park Street: A Publishing Memoir* (Boston: Houghton Mifflin, 1986) focuses on his own era there.

22. Houghton Mifflin, corporate history.

23. John Tebbel, vol. 1 of *History of Book Publishing,* 415.

24. Houghton Mifflin, corporate history.

25. Houghton Mifflin, corporate history.

26. "Story of Houghton Mifflin," 30.

27. Tebbel, *Great Change,* 225.

28. Vivendi sold Houghton Mifflin to the Brookstone Group in 2003.

29. Background on Paul Brooks is drawn from his own books and memoirs and from three interviews I had with him at his home in Lincoln, Mass., in the last few months before he died. In addition to *Two Park Street,* Brooks wrote *The Pursuit of Wilderness* (Boston: Houghton Mifflin, 1971); *The House of Life: Rachel Carson at Work* (Boston: Houghton Mifflin, 1972); *Speaking for Nature: How Literary Naturalists from Henry Thoreau to Rachel Carson Have Shaped America* (Boston: Houghton Mifflin, 1980); *The People of Concord: One Year in the Flowering of New England* (Chester, Conn: Globe Pequot Press, 1990).

30. Brooks, *Two Park Street,* ix.

31. The Brookses were, however, surprised to find themselves in that role, having learned

only after her death that she hoped they would raise the child; PB, interview, 14 July 1997. Also see Linda Lear, *Rachel Carson: Witness for Nature* (New York: Henry Holt, 1997), 481.

32. Awards included the Burroughs Medal from the American Museum of Natural History, the Walter A. Starr Award from the Sierra Club; and the Thoreau Society Medal; and he was named a fellow of the American Academy of Arts and Sciences.

33. Quoted in Brooks, *Two Park Street,* 39.

34. "Story of Houghton Mifflin," 31.

35. Brooks, *Two Park Street,* 24.

36. Lovell Thompson, "Books, Business and Finance," *Publishers' Weekly* 146 (11 November 1944): 1916.

37. Lear, *Carson,* 536 n. 6.

38. Her sister was married to nature author John Kieran; Lear, *Carson,* 536 n. 6.

39. RC to MR, 3 April 1958, HM-HL AM10/3.

40. Quoted in Martha Freeman, ed., *Always, Rachel: The Letters of Rachel Carson and Dorothy Freeman, 1952–1964* (Boston: Beacon, 1994), 352.

41. RC to MR, 3 April 1958, HM-HL AM10/3.

42. RC to MR, 14 April 1958, HM-HL AM10/3.

43. Freeman, *Always,* 394.

44. "Mason" to Mr. Greenstein, 31 March 1962, NYA-NYPL 1292. See also discussion of length in chapter 2.

45. See Gill, *Here at The New Yorker,* 318.

46. Hobbes, personal communication, 27 October 1999.

47. "Rachel Carson's Warning," *New York Times,* 2 June 1962, 28.

48. *The Quoter,* Special Issue 8 (August 1962), NYA-NYPL 1292.

49. All quotes from report of telephone call in memo from Milton Greenstein to "Shawn and Gerdy," 20 June 1962, NYA-NYPL 1292.

50. Milton Greenstein, quoted in obituary, "Milton Greenstein," *New Yorker,* 19 August 1991, 79.

51. Standard Oil Company, advertisement, *New Yorker,* 30 June 1962, 74–75.

52. Freeman, *Always,* 257,

53. Freeman, *Always,* 257.

54. PB, "Report to the Executive Committee," 1 April 1958, RCP-BY 87.

55. PB to RC, 29 March 1960, RCP-BY 87.

56. PB to RC, 13 September 1960, RCP-BY 87. Brooks took considerable pride in having named not only the chapter but the book as well (interview, 14 July 1997). In a letter to Virginia Apgar in November 1962, Carson acknowledged his role: " 'Silent Spring' was suggested by Paul Brooks of Houghton Mifflin. . . . While I have named my other books, it was a long and painful process"; HM-HL AM10/2.

57. PB, interview, 14 July 1997

58. LT to PB, 3 March 1958, RCP-BY 87.

59. PB to RC, 21 December 1959, RCP-BY 87.

60. PB to Carl Buchheister, National Audubon Society, 30 January 1962, RCP-BY 87.

61. PB, interview, 14 July 1997.

62. PB to LT, 15 September 1961, RCP-BY 87.

63. PB to Lois and Louis Darling, 22 December 1961, RCP-BY 42.

64. PB, interview, 14 July 1997.

65. PB to RC, 28 November 1958, RCP-BY 87.

66. PB to RC, 1 December 1959, RCP-BY 87.

67. PB to RC, 18 March 1960, RCP-BY 87.

68. PB to RC, 2 February 1962, RCP-BY 87 (original emphasis).

69. PB to Carl Buchheister, National Audubon Society, 30 January 1962, RCP-BY 87.

70. PB to RC, 2 February 1962, RCP-BY 87.

71. LT, handwritten notes on letter from RC, to PB, 14 February 1959, RCP-BY 87 (original emphasis).

72. Alick Bartholomew to RC, 25 May 1962, RCP-BY 87.

73. PB to RC, 2 February 1962, RCP-BY 87.

74. *Publishers' Weekly* 181 (16 April 1962): 79. Formal announcements in *Publishers' Weekly* 183 (6 August 1962): 72–73; *Kirkus* 130 (15 July 1962): 657; *Library Journal* 87 (15 September 1962): 3059.

75. Personal communication, David McElwain, who had a forty-year career in the educational marketing division, 3 November 1998.

76. AF to MR, 13 April 1962, RCP-BY 87.

77. PB, interview, 14 July 1997.

78. AF to MR, 7 May 1962, RCP-BY 87.

79. LT to MR, 16 July 1962, RCP-BY 105.

80. LT to MR, 26 July 1962, RCP-BY 88.

81. PB to MR, 3 August 1962, RCP-BY 88.

82. Louis A. McLean, Velsicol Corporation, to William E. Spaulding, 2 August 1962, NYA-NYPL 1292; copies also archived in RCP-BY 88 and 89. Velsicol was not among the largest producers of pesticides, but it was the sole producer of chlordane and heptachlor. Although these were not brand-name products, Velsicol's legal position entailed possible harm to the reputation of the chemicals.

83. PB, interview, 2 November 1998.

84. Unsigned memo, 7 August 1962, RCP-BY 88.

85. AF to LT, 19 July 1962, RCP-BY 88.

86. AF to PB, 30 July 1962, RCP-BY 88.

87. AF to Frank Dyckman, marketing consultant, 13 June 1962, RCP-BY 87. His response: "I wouldn't worry too much about agricultural magazines or even companies sabotaging *Silent Spring*. The tobacco company, after all, has not been hurt by all the talk about cancer and the adverse publicity in that case is certainly far worse than what Rachel Carson as one individual has to say"; 18 June 1962, RCP-BY 87.

88. AF to MR, 1 August 1962, RCP-BY 88.

89. AF to PB, 1 August 1962, RCP-BY 88.

90. LT to MR, 21 September 1962, HM-HL AM10/16

91. PB to RC, 8 November 1962, HM-HL AM10/16.

92. AF to MR, 13 April 1962, RCP-BY 87.

93. Houghton Mifflin, advertisement, *San Francisco Sunday Chronicle*, 2 June 1963, 31; RCP-BY 75.

94. *The Story of* Silent Spring, Houghton Mifflin brochure, 1963, draft of copy in RCP-BY 64.

95. PB to RC, 15 November 1962, RCP-BY 88.

96. PB to William S. McCann, M.D., 19 February 1963, HM-HL AM10/16.

97. AF to PB, 26 September 1962, RCP-BY 89.

98. PB, handwritten response to AF on memo, AF to PB, 7 February 1963, RCP-BY 88.

99. AF to Robert Cowan, 30 July 1962, RCP-BY 88.

100. Diane Davin, AF's assistant, to AF, 11 June 1964, RCP-BY 89.

101. AF to Lou Webster, station WEEI, 18 January 1963, RCP-BY 88. Emphasis is Ford's.

102. Houghton Mifflin, advertisement, undated [summer 1962], RCP-BY 75.

103. Brooks, *House of Life*, 294.

104. "The Desolate Year," *Monsanto Magazine* (October 1962): 4–9, RCP-BY 66; reprint in author's possession.

105. See Frank Graham Jr., *Since Silent Spring* (Boston: Houghton Mifflin, 1970), 75.

106. National Agricultural Chemicals Association (NACA), *Fact and Fancy, A Reference Checklist for Evaluating Information about Pesticides*, undated brochure [August 1962], RCP-BY 66; reprint in author's possession.

107. Parke Brinkley, head of NACA, to PB, 1 October 1962, RCP-BY 89.

108. PB, Field Note D-506, 17 October1962, RCP-BY 64. See chapter 4 for further discussion.

109. Houghton Mifflin, *The Story of* Silent Spring, RCP-BY 71.

110. Tom Bethell to PB, 13 January 1963, RCP-BY 64.

111. From handwritten notation on memo from Tom Bethell to PB, 13 January 1963, RCP-BY 64.

112. Postcard, Houghton Mifflin publicity department, undated [March 1963], RCP-BY 75.

113. AF to MR, 12 March and 13 March 1963, RCP-BY 88; Tom Bethell to PB, 22 March 1963, RCP-BY 88.

114. Houghton Mifflin, advertisement, *San Francisco Sunday Chronicle*, 2 June 1963, 31; RCP-BY 75.

115. LT to MR, quoting his own telegram, 4 June 1963, HM-HL AM10/16.

116. PB to MR, 4 June 1963, HM-HL AM10/16.

117. LT to PB, 5 June 1963, HM-HL AM10/16.

118. Ross, *Here but Not Here,* passim.

119. AF to Thomas Horgan, 5 June 1962, RCP-BY 87.

120. AF to MR, 13 April 1962, RCP-BY 87.

121. LT to PB, 3 March 1958, RCP-BY 87.

122. PB to LT, 21 September 1962, RCP-BY 88.

123. William J. Darby, "Silence, Miss Carson," *Chemical and Engineering News* (1 October 1962), 60; reprint, *A Scientist Looks at* Silent Spring, in author's possession.

124. Brooks, *Two Park Street,* 157.

Chapter 4

1. See Frank Graham, *Since Silent Spring* (Boston: Houghton Mifflin, 1970); Craig Waddell, ed., *And No Birds Sing: Rhetorical Analyses of Rachel Carson's* Silent Spring (Carbondale: Southern Illinois University Press, 2000); and Christopher J. Bosso's *Pesticides and Politics* (Pittsburgh: University of Pittsburgh Press, 1987), as well as Linda Lear's biography, *Rachel Carson: Witness for Nature* (New York: Henry Holt, 1997).

2. See Graham, Lear, and also Lear's articles "Bombshell in Beltsville: The USDA and the Challenge of *Silent Spring*," *Agricultural History* 66 (Spring 1992): 151–70, and "Rachel Carson's *Silent Spring*," *Environmental History Review* 17 (Summer 1993): 23–48.

3. The Food and Drug Administration and the Public Health Service were also involved, as they had been historically; but their public involvement with the *Silent Spring*–era discussion was somewhat less than USDA's until congressional hearings began.

4. Graham, *Since Silent Spring*, 39.

5. I. L. Baldwin, "Chemicals and Pests," *Science*, 28 September 1962, 1043; reprint in author's possession.

6. George Decker, "Pros and Cons of Pests, Pest Control and Pesticides," *World Review of Pest Control* 1 (spring 1962): 6–18; reprint in author's possession.

7. National Academy of Sciences-National Research Council, *Pest Control and Wildlife Relationships* (Washington, D.C.: Government Printing Office, 1962, 1963).

8. Quoted in Graham, *Since Silent Spring*, 42.

9. See Ernest G. Moore, *The Agricultural Research Service* (New York: Praeger, 1967), esp. 189–207.

10. Byron T. Shaw to Orville Freeman, 11 July 1962, NARA RG 16, 3782.

11. Rodney Leonard to Orville Freeman, 12 July 1962, NARA RG 16, 3782.

12. Correspondence, NARA RG 16, 3782.

13. *Wall Street Journal*, 3 August 1962, 1 (blurb in "Short-Takes" section).

14. Correspondence, NARA RG 16, 3959.

15. "Are Weed Killers, Bug Sprays Poisoning the Country?" *U.S. News & World Report*, 26 November 1962, 86–94.

16. E. Bruce Harrison, *Going Green: How to Communicate Your Company's Environmental Commitment* (Burr Ridge, Ill.: Irwin, 1993), xv. According to *Barron's*, at the time "virtually all the major chemical firms turn out pesticides. Besides Union Carbide, the list includes DuPont, Dow, Allied Chemical, American Cyanamid, Monsanto, Olin Mathieson and Hercules Powder. Several major oil companies, notably Shell and Standard of California, also are sizable producers. In fact, a subsidiary of the latter, California Chemical Co., reportedly is the world's largest maker of pesticides"; "No Year of the Locust," *Barron's*, 18 June 1962, 11. Thus Harrison's list of corporations most actively cooperating in the public relations campaign was complete suggests that not all pesticide producers, nor perhaps even the largest, were involved in the planned opposition to *Silent Spring*.

17. Thomas H. Jukes, interviewed in film "Rachel Carson's *Silent Spring*," produced by Neil Goodwin, *The American Experience*, Public Broadcasting System, ©The History Consortium, 1992; ©WGBH Educational Foundation, WNET/Thirteen, and Peace River Films, 1993.

18. Thomas H. Jukes, "A 'Balance of Nature' Story As It Might Have Been," *NAC News and Pesticide Review*, August 1962, 15.

19. "The Desolate Year," *Monsanto Magazine* (October 1962): 4–9, RCP-BY 66; reprint in author's possession.

20. Dan J. Forrestal, *Faith, Hope and $5,000: The Story of Monsanto* (New York: Simon and Schuster, 1977), 195.

21. "Industry Maps Defense to Pesticide Criticisms," *Chemical & Engineering News* (*C&EN*), 13 August 1962, 24.

22. Public statement, Allied Chemical Company, August 1962, RCP-BY 66.

23. Stare wrote regularly on the subject for the American Medical Association magazine, *Today's Health*.

24. "How Do You Fight a Best-Seller?" *Printers' Ink* (30 November 1962), 42.

25. According to *Printers' Ink*, the remaining 20 percent of retail sales represented export, most logically for foreign agriculture. "How Do You Fight a Best-Seller?" 42.

26. "Feeling Little Pain," *Business Week*, 25 May 1963, 36.

27. Data cited, among many other places, in Robert C. Cowen, "The Chemical War on Pests: Is It Getting Out of Hand?" *Christian Science Monitor*, 10 August 1962, 9; "How Do You Fight a Best Seller?"; and in Lawrence Galton, "Great Debate Over Pests—And Pesticides," *New York Times Sunday Magazine*, 14 April 1963, 34.

28. U.S. Department of Commerce, *Census of Manufactures 1963*, vol. 1, *Summary and Subject Statistics*, and vol. 2, *Industry Statistics*. Also see *Census of Business 1963, Summary Wholesale Statistics* (Washington, D.C.: U.S. Government Printing Office, 1996).

29. "Feeling Little Pain," 36.

30. "How Do You Fight a Best-Seller?" 42.

31. Several such ads may be found in DuPont's archives, PR/DNI—HM&L.

32. Deeper discussion of the industry's public-relations efforts is found in Gerald Markowitz and David Rosner, *Deceit and Denial: The Deadly Politics of Industrial Pollution* (Berkeley: University of California Press, 2002), 139–67.

33. See Markowitz and Rosner, *Deceit and Denial*, 142.

34. "Industry Maps Defense," *C&EN*, 25.

35. "Feeling Little Pain," *Business Week*, 36.

36. "More of Rachel," editorial, *Agrichemical West* (October 1962): 24.

37. Allan E. Settle, quoted in Harrison, *Going Green*, xiv. Also see Forrestal, *Faith, Hope and $5,000*, 194–97.

38. Quoted in Graham, *Since Silent Spring*, 65.

39. "Industry Maps Defense," *C&EN*, 24.

40. "How Do You Fight a Best-Seller?" 42. See also "Industry Maps Defense," 24.

41. Cecil H. Chilton, "Let's Not Underestimate the Power of Women," *Chemical Engineering* (1 October 62): 7.

42. Charles Sommer, quoted in Forrestal, *Faith, Hope and $5,000*, 196.

43. National Agricultural Chemicals Association (NACA), *Fact and Fancy, A Reference Checklist for Evaluating Information about Pesticides*, undated brochure [August 1962]. RCP-BY 66; reprint in author's possession.

44. "The Desolate Year," *Monsanto Magazine*. (October 1962): 4–9; reprint in author's possession. The reference to malaria is noteworthy. The World Health Organization had offered some early objections to the *New Yorker* articles because of its anti-malaria campaign, thus, corporate opposition could well have hoped to draw on WHO as an authoritative ally. See correspondence and "Notice of Meetings," National Citizens Committee for WHO, RCP-BY 88. Carson noted the inconsistency of WHO, which earlier had studied and reported on dangers of pesticide abuse and residues.

45. Louis McLean, *The Necessity, Value, and Safety of Pesticides*, (n.p.) in author's possession.

46. Darby, "Silence, Miss Carson," *C&EN*, 1 October 1962, 60; reprint, *A Scientist Looks at Silent Spring*, in author's possession.

47. Frederick J. Stare, "Some Comments on *Silent Spring*," from *Nutrition Reviews* (January 1963), reprint in author's possession.

48. I. L. Baldwin, "Chemicals and Pests," *Science*, 28 September 1962, 1042.

49. New York State College of Agriculture, Ithaca, and New York State Agricultural Experiment Station, Geneva, *Facts on the Use of Pesticides,* brochure, October 1962, in the author's possession.

50. Decker, "Pros and Cons."

51. "Sound Advice to Book Readers," editorial from *Progressive Farmer*, excerpted in *NAC News and Pesticide Review* (December 1962): 2. The comment on thalidomide was also a potshot at the Food and Drug Administration, which stood to gain clout if agency oversight of pesticides were reorganized.

52. Baldwin, "Chemicals and Pests," 1043.

53. "Feeling Little Pain," *Business Week*, 36.

54. "Industry Maps Defense," *C&EN*, 24.

55. Stare, "Some Comments," 1.

56. William B. Bean, "The Noise of *Silent Spring*," *Archives of Internal Medicine* 112 (September 1963): 309. This journal is a publication of the American Medical Association, within which Stare had some standing.

57. Darby, "Silence, Miss Carson," 61.

58. C. G. King, President, The Nutrition Foundation, Inc., cover letter to mass mailing, January 1963; in author's possession.

59. E. M. Adams, quoted in John M. Lee, "*Silent Spring* Is Now Noisy Summer," *New York Times*, 22 July 1962, II, 11.

60. Sidebar to Lorus and Margery Milne, "There's Poison All Around Us Now," review, *New York Times Book Review*, 23 September 1962, 26.

61. Stare, "Some Comments," 3.

62. Quoted in "NACA Readies Counterattack on Carson's *Silent Spring*, PR Job Will Be a Big One," *Oil, Paint and Drug Reporter* (10 September 1962), PR/DNI—HM&L.

63. White-Stevens interviewed on *CBS Reports*, "The Silent Spring of Rachel Carson," broadcast 3 April 1963, produced and written by Jay L. McMullen, © Columbia Broadcasting System.

64. For example, in Decker's "Pros and Cons," 1.

65. NACA, *Fact and Fancy*, 1.

66. "Monsanto Chemical Company Published a Rebuttal to Rachel Carson's *Silent Spring*," *PR News* (February 1963), reprinted in *PR News Casebook: 1000 Public Relations Case Studies*, ed. David P. Bianco (Detroit: Gale Research, 1993), 440.

67. Syndicated columnist Elmer Roessner called the parody "much calmer, much more judicial . . . far less emotional" than *Silent Spring*; quoted in "Naturalist-Author's New Book Worries Chemical Industry," *Fort Wayne (Ind.) News Sentinel*, 3 October 1962, PR/DNI—HM&L.

68. "The Day of the Locust: Monsanto's 'Desolate Year'," *Printers' Ink* (30 November 1962): 42.

69. Lee, "*Silent Spring* Is Now Noisy Summer," 11.

70. Stare's name never appeared without all four degrees noted, "B.S., M.S., Ph.D., M.D.," as well as his affiliation with Harvard.

71. Lear, *Carson*, 429.

72. Bean, "Noise of *Silent Spring*," 311.

73. See Mary A. McCay, *Rachel Carson* (New York: Twayne, 1993), and H. Patricia Hynes, *The Recurrent Silent Spring* (New York: Pergamon, 1989), as well as Lear on the issue.

74. Lear, *Carson*, 436, 463.

75. See Lear, *Carson*, 435–38, regarding Westcott's ties to industry.

76. King, cover letter.

77. AF to Lou Webster, radio station WEEI, 29 January 1962, RCP-BY 88.

78. Louis A. McLean, Velsicol Corporation, to William E. Spaulding, 2 August 1962, NYA-NYPL 1292; copies also archived in RCP-BY 88 and 89.

79. McLean, *Necessity, Value, and Safety*, 14.

80. McLean, *Necessity, Value, and Safety*, 17.

81. Forrestal, *Faith, Hope and $5,000*, 194.

82. Parke C. Brinkley to Rodney Leonard, 29 August 1962, NARA RG 16, 3959.

83. "Industry Maps Defense," 25.

84. Public Relations Department, DuPont Nemours, "Press Analysis for September 5, 1962," PR/DNI—HM&L.

85. "How Do You Fight a Best-Seller?" 43.

86. "Industry Maps Defense," 24.

87. See correspondence between LT and MR: 16 July 1962, RCP-BY 105; 26 July 1962, RCP-BY 88.

88. LT to PB, 8 January 1963, HM-HL AM 10, box 16. The neighbor recommended that Houghton Mifflin take out a large ad for Carson in the widely read Harvard student paper, *The Crimson*, in order to reach "pretty directly most of the people that Stare is likely to influence with his local hate campaign."

89. "Monsanto . . . Rebuttal," 440.

90. "How Do You Fight a Best-Seller?" 43.

91. Roger B. May, "Charges That Chemicals Harm Humans, Nature Fails to Hamper Sales," *Wall Street Journal* 3 April 1963, 16.

92. PB to RC, 15 November 1962, RCP-BY 88.

93. National Pest Control Association service letter, "Rachel Carson Writes on Pesticides," 22 August 1962, RCP-BY 66.

94. *County Agent and Vo-Ag Teacher*, "How to Answer Rachel Carson," RCP-BY 66.

95. Byron T. Shaw, USDA, to the secretary, 14 August 1962, NARA RG 16, 3782.

96. "Industry Maps Defense," 25.

97. Sidebar to Milne, "There's Poison All Around Us Now," *New York Times Book Review*, 26.

98. "Monsanto . . . Rebuttal," 440. One such galley proof was included in the Carson archives, suggesting that someone had made sure she or Houghton Mifflin had early warning about the parody.

99. "How Do You Fight a Best-Seller?" *Printers' Ink*, 43.

100. "Monsanto Dissects Pesticide Criticism," *New York Times*, 22 September 1962, 29.

101. Ernest Moore, USDA, to Harland Manchester, 3 October 1962, NARA RG 310, 651.

102. Ernest Moore, USDA, to R. Milton Carleton, 9 November 1962, NARA RG 310, 651.

103. AF to PB, 5 February 1963, RCP-BY 88.

104. This meeting was covered nationally by wire services and in syndicated *New York Times* coverage by Walter Sullivan; "Chemists Debate Pesticides Book," *New York Times*, 13 September 1962, 34.

105. Walter Sullivan, "Books of the Times," *New York Times*, 27 September 1962, 35.

106. Harriet van Horne reported the plans in her syndicated column for 30 August; *New York World-Telegram and Sun*, 30 August 1962; RCP-BY 62.

107. "Not a Pesticide Scare, Please!" *American Fruit Grower*, October 1962, PR/DNI—IIM&L.

108. Jay McMullen, interviewed in "Rachel Carson's *Silent Spring*," *American Experience*.

109. See Val Adams, "2 Sponsors Quit Pesticide Show," *New York Times*, 3 April 1963, 95. A Kiwi Polish executive later wrote to CBS's Fred Friendly, with a copy to Paul Brooks at Houghton Mifflin, expressing the company's pleasure in being associated with the program and noting that the firm had received nothing but praise for having sponsored it. "I might say at this point that in our 15 years of sponsorships and various advertising presentations in the U.S., we have never had such a fine reaction. It is certainly my feeling that for everyone who writes, thousands feel the same way, but will not take the time to do so." Lawrence Emley to Fred Friendly, 25 May 1963, RCP-BY 89.

110. Lysol's ad was switched to *Mr. Ed*.

111. Warner Twyford, "3 Sponsors Won't Sing of Spring," *Norfolk (Va.) Virginian-Pilot*, 3 April 1963, RCP-BY 75.

112. "CBS Reports' Sponsors Drop Wednesday Segment," *Broadcasting* (1 April 1963): 10.

113. Jukes, interviewed in "Rachel Carson's *Silent Spring*," *American Experience*.

Chapter 5

1. "Hiss of Doom?" *Newsweek*, 6 August 1962, 55. The article appeared three pages after a report on Sherri Finkbein's legal battles with respect to thalidomide-related birth defects; "I Pray They'll Hurry," *Newsweek*, 6 August 1962, 52.

2. See brief discussion of Bourdieu in the appendix.

3. *Omaha World Herald*, 23 September 1962, review, PR/DNI—HM&L.

4. "Could Be Year's Most Noted Book," *Cincinnati Enquirer*, 29 September 1962, 10.

5. Albert B. Southwick, "Scientists with Something to Say," review, *Worcester (Mass.) Telegram*, 7 October 1962, RCP-BY 62.

6. Harry Nelson, "Mosquito Sterilizer Tried Out," *Los Angeles Times*, 23 September 1962, 6.

7. William K. Wyant, "Bug and Weed Killers; Blessings or Blights?" editorial, *St. Louis Post-Dispatch*, 29 July 1962," 1.

8. Miles A. Smith, Associated Press, "Frightening Tocsin Tolled by Biologist," *Indianapolis News*, 6 October 1962, 2.

9. Elmer Roessner, "Naturalist-Author's New Book Worries Chemical Industry," *Fort Wayne (Ind.) News Sentinel*, 3 October 1962, archives, PR/DNI—HM&L. The article was syndicated, and other papers applied other headlines, for example, the *Norfolk Virginian-Pilot*'s "*Silent Spring* Strikes Business Alarm Bell."

10. The term originally appeared in John Lee's July report in the *Times* but was quoted repeatedly for months thereafter; John M. Lee, "*Silent Spring* Is Now Noisy Summer," *New York Times*, 22 July 1962, III, 1.

11. George Bragdon, *Hartford Times*, 1 November 1961, PR/DNI—HM&L.

12. John Lear, "Personality Portrait," *Saturday Review*, 1 June 1963, 45.

13. Barbara Yuncker, "A Voice Amid the Silence," *New York Post*, 30 September 1962, 2, but also run in other regional and local papers.

14. Ann Cottrell Free's profile appeared as part of her syndicated series in early January 1963 in the *Des Moines Register* and in February in the *Louisville Courier-Journal*, RCP-BY 88 and 108.

15. For an excellent historical discussion of the journalistic ideal of objectivity, see Michael Schudson, *Discovering the News: A Social History of American Newspapers* (New York: Basic, 1978). See also Dan Schiller, *Objectivity and the News: The Public and the Rise of Commercial Journalism* (Philadelphia: University of Pennsylvania Press, 1982); Mitchell Stephens, *A History of the News: From the Drum to the Satellite* (New York: Viking, 1988); and Hazel Dicken-Garcia, *Journalistic Standards in Nineteenth-Century America* (Madison: University of Wisconsin Press, 1989).

16. Comparative data were used frequently in White-Stevens's speeches; see Associated Press story on such a speech, *Ann Arbor (Mich.) News*, 12 December 1962, RCP-BY 61.

17. Press release, CBS News, "CBS Reports Proves Perils and Promise in Widespread Pesticide Use April 3," 14 March 1963; NARA RG 16, 3959.

18. For one example, see Walter Hawver, "Rachel Carson Holds Her Own," *Albany (N.Y.) Knickerbocker News*, 4 April 1963, RCP-BY 75.

19. E.g., Marilyn Aycock, "Where Viewers Need Facts, *Silent Spring* Is Silent," *Louisville Courier-Journal*, 4 April 1963, RCP-BY 75.

20. See further discussion of Habermas's "public sphere" in the appendix.

21. See a discussion of book reviewing in the appendix. Also, regarding the *New York Times* in particular John Gross, "The 'Littery Supplement' Comes of Age: A History, of Sorts, of the Book Review," *New York Times Book Review*, 6 October 1996, 9–11.

22. Oliver LaFarge, "The Carson Book," *Santa Fe New Mexican*, 7 October 1962, PR/DNI—HM&L.

23. Tom Mulvaney, *Houston Chronicle*, to AF, letter, 26 November 1962, RCP-BY 88.

24. Edward P. Ryan, "Weed Killers or Man Killers?" *Boston Pilot*, 29 September 1962, RCP-BY 62.

25. "Are We Poisoning Ourselves?" *Business Week*, 8 September 1962, 26.

26. Sidebar with review of *Silent Spring, Richmond News Leader*, 20 September 1962, PR/DNI—HM&L.

27. The Rachel Carson Papers occupy 117 boxes in 53.5 linear feet of the Yale Collection in American Literature at the Beinecke Library.

28. Eric Sevareid, "Pests vs. Men: The Big Battle Is Raging Again," *Los Angeles Times*, 23 September 1962, G2. This column, like all of Sevareid's, was syndicated, and therefore it

appeared in many papers at various times roughly between 22 September and 9 October 1962, when it appeared in the *Washington Star*.

29. Phil Robertson, review, *Hollister (Calif.) Free Lance*, RCP-BY 62.

30. Harriet van Horne, "Crop-Dusting Documentary Is Aired by ABC," review of program "ABC, 'Focus on America,' " *New York World-Telegram and Sun*, 30 August 1962, RCP-BY 62.

31. Lorus and Margery Milne, "There's Poison All Around Us Now," *New York Times Book Review*, 23 September 1962, 1.

32. Walter Sullivan, "Books of the Times," *New York Times*, 27 September 1962, 35.

33. Lawrence Galton, "Great Debate Over Pests—And Pesticides," *New York Times Sunday Magazine*, 14 April 1963, 34. 80–82.

34. "Pesticide, Cranberry Scares Are 'Baloney'," *New York Herald Tribune* Service, 11 September 1962, PR/DNI—HM&L.

35. White-Stevens had misused Audubon Society statistics in an effort to prove the harmlessness of pesticides to birds, and Smith called him on it. Art Smith, "Fishing and Hunting," *New York Herald Tribune*, 11 April 1963, 25.

36. Arline Grimes, "Authoress Tells Bostonians About Chemical Dangers," *Boston Herald*, 18 January 1963, RCP-BY 66.

37. AF to PB, memo, 5 February 1963, RCP-BY 88.

38. Carson, *Silent Spring*, 58.

39. Howard James, "Elm Disease Spraying Irks Bird Lovers," *Chicago Tribune*, 1 July 1962, 32.

40. "Worse Than Insects?" *Time*, 11 April 1949, 70.

41. The feud is described by John Tebbel and Mary Ellen Zuckerman in *The Magazine in America, 1740–1990* (New York: Oxford, 1991), 220–21. *Time* had written a "disdainful" review of the *New Yorker's* first issue in 1925. Ross assigned writer Wolcott Gibbs to create a profile of Luce parodying *Time's* writing style, which according to Tebbel and Zuckerman, "Luce neither forgot nor forgave. . . . Years later, still haunted, he referred to 'that goddam article.' "

42. "Pesticides: The Price for Progress," *Time*, 28 September 1962, 45–47.

43. N. W. Ayer's report on 1963 magazine circulations lists *Time* at 2.77 million, *Newsweek* at 1.53 million, and *USN&WR* at 1.26 million; *N. W. Ayer & Son's Directory of Newspapers and Periodicals* (Philadelphia: N. W. Ayer & Son, 1963), 1397–1417.

44. "Are Weed Killers, Bug Sprays Poisoning the Country?" *US News & World Report (USN&WR)*, 26 November 1962, 86–94.

45. "How Much Danger from Pesticides?" *USN&WR*, 27 May 1963, 8.

46. "If You Didn't Have Poison Sprays," *USN&WR*, 3 June 1963, 74.

47. "How Much Danger?" *USN&WR*, 8.

48. "Psychiatry: Insects' Revenge," *Newsweek*, 24 July 1961, 48.

49. "Hiss of Doom?" 55.

50. "Judgment on Pesticides," *Newsweek*, 27 May 1963, 69.

51. "Pests and Poisons," *Newsweek*, 17 June 1963, 86.

52. Ayer lists *Reader's Digest's* 1963 circulation at 13.61 million; *TV Guide* ranked second at 8.17 million. Other magazines ranked by circulation were (with circulation in millions) *McCall's* (8.14), *Look* (7.14), *Ladies' Home Journal* (7.14), *Life* (7.07), *Saturday Evening Post* (6.63), *Family Circle* (6.53), *Better Homes & Gardens* (5.9), *Women's Day* (5.6), *Time* (2.77), *Redbook* (2.53), *Newsweek* (1.53), and *USN&WR* (1.26); *Directory of Newspapers and Periodicals*, 1397–1417.

53. Reported in Tebbel and Zuckerman, *Magazine in America*, 184–85.

54. RC to DeWitt Wallace, 27 January 1958, RCP-BY 42; Walter B. Mahony to RC, 30 January 1958, RCP-BY 42.

55. Robert S. Strother, "Backfire in the War Against Insects," *Reader's Digest*, July 1959, 64–69.

56. "Good-By to Garden Pests?" *Reader's Digest*, May 1961, 158–62.

57. PB to RC, letter, 18 October 1962, RCP-BY 88.

58. "Are We Poisoning Ourselves With Pesticides?" *Reader's Digest*, December 1962, 51.

59. John Strohm and Cliff Ganschow, "The Great Pesticide Controversy," *Reader's Digest*, October 1963, 12–28.

60. "War Against the Insects," *Life*, 11 May 1962, 74–76.

61. "Gentle Storm Center," *Life*, 12 October 1962, 105–10.

62. "A disagreement over how to proceed ended the collaboration" was the remainder of the caption; Edwin Diamond, "The Myth of the 'Pesticide Menace'," *Saturday Evening Post*, 28 September 1963, 16–17.

63. I. L. Baldwin, "Chemicals and Pests," *Science*, 28 September 1962, 1042–43.

64. PB to MR, memo, 22 October 1962, HM-HL AM10/16.

65. William Vogt, "Reviews: On Man the Destroyer," *Natural History*, January 1963, 3.

66. LaMont C. Cole, "Books: Rachel Carson's Indictment of the Wide Use of Pesticides," review, *Scientific American*, December 1962, 174–78.

67. "Are We Poisoning Ourselves?" *Business Week*, 8 September 1962, 36.

68. Business and trade journals often referred to the PSAC panel and report as the "Wiesner" panel and report, highlighting the person of committee chair Jerome Wiesner, a Kennedy appointee. Text and photos also spotlighted Wiesner rather than the composition of the panel.

69. Richard C. Davids, "You're Accused of Poisoning Food," *Farm Journal*, September 1962, 29 ff.

70. Robert A. Kehoe, review, *Farm Magazine*, Spring 1963, 29 ff.

71. B. S. Wright, "The Plight of the Woodcock: Dieldrin and Heptachlor," *Field & Stream*, September 1962, 34–35.

72. Virginia Kraft, "The Life-Giving Spray," *Sports Illustrated*, 18 November 1963, 24–26.

73. "Pesticides: Attack and Counterattack," *Consumer Reports*, January 1963, 37–39.

74. "The Public Needs to Be Alarmed," *Consumer Reports*, July 1963, 324–26; and "The Proper Use of Home Insecticides," *Consumer Reports*, August 1963, 392–93.

75. Hamilton Mason, "Should Gardeners Stop Dusting and Spraying?" editorial, *Better Homes and Gardens*, November 1962, 6.

76. "The Better Way: The Pesticide Dangers You Really Face," *Good Housekeeping*, June 1963, 147–49.

77. "Better Way," 147.

78. "Spotting the Food Quack," *Today's Health*, September 1962, 11.

79. Howard Earle, "Pesticides: Facts, Not Fear," *Today's Health*, February 1963, 19 ff.

80. Robert Rodale, "*Silent Spring*: Rachel Carson's Masterpiece," *Prevention*, October 1962, 45 ff.

81. His manuscript is included in the Yale archives with a cover letter to Carson, RCP-BY 62.

82. "James Rorty, "Varieties of Poison," review, *Commonweal*, 14 December 1962, 320–21.

83. Marston Bates, "Man and Other Pests," *Nation*, October 1962, 202.

84. Edward Weeks, "The Peripatetic Reviewer," review, *Atlantic*, October 1962, 134–35. Directly opposite Weeks's review was a full-page advertisement from Houghton Mifflin, "The Peripatetic Advertiser," in paragraphed-text form, featuring *Silent Spring*.

85. Clark Van Fleet, "*Silent Spring* on the Pacific Slope: A Postscript to Rachel Carson," *Atlantic*, July 1963, 81–84.

86. See memo, AF to PB, 30 July 1962, RCP-BY 88.

87. William O. Douglas, "Dissent in Favor of Man," *Saturday Review*, 5 May 1960, 59–60.

88. "J. L." (John Lear), "Patroness of Birdsong," *Saturday Review*, 1 June 1963, 45.

89. Loren Eiseley, "Using a Plague to Fight a Plague," *Saturday Review*, 29 September 1962, 18 ff.

90. "Buzz, Buzz, Buzz," *New Republic*, 13 August 1962, 7.

91. "Those Pesticides," *New Republic*, 17 August 1963, 6.

92. S. C., Hamden, Conn., to RC, 4 April 1963, RCP-BY 89.

93. Despite that provision, Carson did appear on short notice on the NBC *Today* show a few days before the PSAC report was made public, where she was interviewed by Hugh Downs; see correspondence, RC to AF, 12 May 1963, RCP-BY 89.

94. AF to RC, 7 November 1963, RCP-BY 88, concerning "The Hidden Menace."

95. Van Horne, "Crop-Dusting Documentary Is Aired by ABC."

96. National Film Board of Canada, *Poisons, Pests, and People*, produced by David Bairslow, © 1960.

97. "Educational TV WEDH Channel 24—Special Program," *Meriden (Conn.) Journal*, 7 November 1962, RCP-BY 101.

98. *CBS Reports*, "The Silent Spring of Rachel Carson," broadcast 3 April 1963, produced and written by Jay L. McMullen, © Columbia Broadcasting System, 1963.

99. "Close-up: CBS Reports—Documentary," *TV Guide*, vol. 3, 30 March 1963 (New York edition), A-59; RCP-BY 75.

100. Val Adams, "2 Sponsors Quit Pesticide Show," *New York Times*, 3 April 1963, 95. CBS ads for its program were located on the same page.

101. Transcript of *CBS Reports*, "The Silent Spring of Rachel Carson," RCP-BY 75.

102. The program may have been sidelined during some administration shuffling at CBS.

103. *CBS Reports*, "The Silent Spring of Rachel Carson."

104. *CBS Reports*, "The Verdict on the Silent Spring of Rachel Carson," 15 May 1963 broadcast, written and produced by Jay L. McMullen, © Columbia Broadcasting System, 1963, transcript in RCP-BY 75.

105. "For Many a Spring," *Time*, 24 April 1964, 73.

106. Glendy Culligan, *Washington Post*, telegram to AF, 16 May 1963; RCP-BY 89.

107. Brooks Atkinson to AF, 17 July 1962, RCP-BY 88.

108. Brooks Atkinson, "Critic at Large," *New York Times*, 11 September 1962, 30.

109. Harry Hansen, "A Book Not Out Yet Reaping Praise and Condemnation," *Chicago Tribune*, 23 September 1962, IV, 7. Hansen, however, wrote for the *New York Herald Tribune*, and it is likely that this originated from there.

110. See Gerald Howard, "The Cultural Ecology of Book Reviewing," *Media Studies Journal* 6 (Summer 1992); 95; also discussed in the conclusion.

Chapter 6

1. H. P., Houghton Mifflin, to RC, 27 April 1962, RCP-BY 87.

2. See chapter 5; S. C., Hamden, Conn., to RC, 4 April 1963, RCP-BY 89.

3. In 1972 *New York Times* letters-page editor Kalman Seigel wrote, "We try to give voice to a variety of positions and issues crying for exposure, to opinions that differ with the paper's editorial stand and to rebuttals and dialogue with previously published letter writers. . . . On highly controversial subjects where the mail on one side is preponderantly heavy, the ratio used gives a clue to, not an exact reflection of, the numbers received on each side.

On highly controversial issues on which the paper has taken a stand, weight is given to positions opposed to the *Times'* editorial policy." Kalman Seigel, *Talking Back to the New York Times* (New York: Quadrangle, 1972), 9.

4. Letters, *Time,* 5 October 1962, 9.

5. Letters, *Sports Illustrated,* 2 December 1962, 96.

6. Letters, *Life,* 2 November 1962, 21.

7. Edwin Diamond, "The Myth of the 'Pesticide Menace'," *Saturday Evening Post,* 28 September 1963, 16–17.

8. Letters, *Saturday Evening Post,* 26 October 1963, 4.

9. Richard L. Kenyon, "Assault on Nature," editorial, *Chemical and Engineering News* (*C&EN*), 23 July 1962, 5; and "Pesticides on Trial," editorial, *C&EN,* 30 July 1962, 5.

10. Richard L. Kenyon, "A Problem in Communications," editorial, *C&EN,* 15 October 1962, 5.

11. One letter appeared in the *Des Moines Register* on 15 July 1962 and two on 22 July 1962, concurrent with an editorial against roadside spraying.

12. "Editor's Note" following letter from Charles D. Carleton to the *Memphis (Tenn.) Commercial Appeal,* 30 September 1962, RCP-BY 71.

13. William J. Lederer, *A Nation of Sheep* (New York: Norton, 1961).

14. Lederer, *Nation of Sheep,* 7–8.

15. Lederer, *Nation of Sheep,* passim, with passages consolidated into this single quote in a Western Electric corporate condensation for the Western Electric Booklet Rack Service for Employees series, n.d. but ca. 1963, NYA-NYPL 1292.

16. Seigel, *Talking Back,* 6.

17. Seigel, *Talking Back,* 6–7. He was referring to a study of letter writers to three Kansas City newspapers: Gary L. Vacin, "A Study of Letter-Writers," *Journalism Quarterly* 42 (summer 1965): 464–65, 510.

18. For quantitative analysis of the letters to the *New Yorker,* see app. N in Priscilla Coit Murphy, "What a Book Can Do: *Silent Spring* and Media-Borne Public Debate" (PhD. Diss., University of North Carolina at Chapel Hill, 2000).

19. Although 75 percent of the magazine's circulation came from outside the metropolitan area, the magazine was most heavily read on the East Coast, particularly in suburban areas of Boston, Philadelphia, and Washington.

20. E. H., Washington, D.C., letter to *The New Yorker,* 29 June 1962, NYA-NYPL 1292.

21. Mrs. Paul Schulz, letter to the editor, *Des Moines Register,* 29 July 1962, 15-G.

22. Edward Misiaszek, letter to the editor, *Norwich (Conn.) Bulletin,* 17 September 1962, RCP-BY 71.

23. Lon Campbell, "Waters and Woods: The Pros and Cons of Pesticides, Herbicides," *Toledo (Ohio) Times,* 3 April 1962, RCP-BY 89.

24. Jeanne Goode, letter to the editor, *Riverdale (N.Y.) Press,* 1 November 1962, RCP-BY 71.

25. F. D., Louisville, KY., letter to *The New Yorker,* 4 July 1962, NYA-NYPL 1292.

26. E. S., Amherst, Mass., letter to *The New Yorker,* 30 June 1972, NYA-NYPL 1292.

27. Herbert Insley, letter to the editor, *C&EN,* 20 August 1962, 5.

28. Mrs. T. G., Urbana, Ill., letter to *The New Yorker,* 2 July 1962, NYA-NYPL 1292.

29. L. C., London, Eng., letter to *The New Yorker,* n.d., NYA-NYPL 1292.

30. Mrs. Robert Jones, letter to the editor, *Dallas News,* 15 October 1962, RCP-BY 71.

31. Bruno E. Schiffleger, letter to the editor, *Milwaukee Journal,* 30 November 1962, RCP-BY 70.

32. D. S., Topeka, Kan., letter to *The New Yorker,* 10 July 1962.

33. Basil S. Farah, letter to the editor, *C&EN,* 22 October 1962, 5.

34. The ad appeared in the 25 November issue of the *New York Times Book Review,* 64,

and drew some bemused attention at Houghton Mifflin: "Tom Denny of the *Times* tells me it is a magazine owned by a man or two men who do 'good, cheerful deeds.' When they come across something they feel will do good, they run the kind of ad that appeared in the *Times*. They hope in spreading the word, they in turn will get an advertisement in their magazine *Free Deeds*. . . . Sounds like 'lighting a candle' or 'a good deed in a naughty world.' " Memo, Catherine Lewis to PB, 13 December 1962, RCP-BY 88.

35. Mary W. Smelker, letter to the editor, *Washington Post*, 9 July 1962, A14.

36. Mrs. Charles Blades, letter to the editor, *Topeka (Kans.) State Journal*, 15 February 1963, copy to RC, RCP-BY 61.

37. A. Scott Walker, letter to the editor, *Lakeville (Conn.) Journal*, 4 October 1962, RCP-BY 71.

38. Mrs. Vicki Patterson, letter to the editor, *Time*, 5 October 1962, 9.

39. W. B., Farina, Ill., letter to *The New Yorker*, n.d., NYA-NYPL 1292.

40. See memos of correspondence, 10 January 1963, RCP-BY 88.

41. Kenneth Birkhead, Assistant to the Secretary, to Mrs. R. B., Onekama, Mich., 24 September 1962, NARA RG 16, 3782.

42. C. B., Hartsville, Pa., to Orville Freeman, 5 April 1963, NARA RG 310, 651.

43. D. R., Washington, D.C., to Orville Freeman, 3 April 1963, NARA RG 310, 651.

44. Murphy, "What a Book Can Do," app. N.

45. As of December 1962, Houghton Mifflin had printed 500,000 copies.

46. R. H., Mill Valley, Calif., letter to *The New Yorker*, 29 June 1962, NYA-NYPL 1292.

47. Charles C. Coutant, letter to the editor, *Bethlehem (Pa.) Globe-Times*, 7 September 1962, RCP-BY 71.

48. R.C., speech, Women's National Press Club, 5 December 1962, RCP-BY 101.

49. Cecile Young, letter to the editor, *Rochester (N.Y.) Democrat and Chronicle*, 20 January 1963, RCP-BY 90.

50. Mrs. Charles Blades, letter to the editor, *Topeka (Kans.) State Journal*, 15 February 1963, RCP-BY 61.

51. Charles C. Coutant, letter to the editor, *Bethlehem (Pa.) Globe-Times*, 7 September 1962, RCP-BY 71.

52. Seigel, *Talking Back*, 9.

53. Allen Klein, letter to the editor, *New York Post Magazine*, 1 October 1962, 4.

54. Joseph Hoffman, letter to the editor, *C&EN*, 22 October 1962, 5.

Conclusion

1. Oscar H. Gandy Jr., *Beyond Agenda Setting: Information Subsidies and Public Policy* (Norwood, N.J.: Ablix, 1982).

2. Gerald Howard, "The Cultural Ecology of Book Reviewing," *Media Studies Journal* 6 (Summer 1992): 95.

3. Scott Slovic, *Seeking Awareness in American Nature Writing: Henry Thoreau, Annie Dillard, Edward Abbey, Wendell Berry, Barry Lopez* (Salt Lake City: University of Utah Press, 1992), 169.

4. Eliot Weinberger, in "Symposium: Twelve Visions," ed. C. Barber, *Media Studies Journal* (Summer 1992): 42.

5. See chap. 6, "The Transformation of the Public Sphere's Political Function," esp. pp. 188–94, in Jürgen Habermas's 1962 *The Structural Transformation of the Public Sphere: An Inquiry into a Category of Bourgeois Society* (Cambridge: MIT Press, 2001), and his later "Further Reflections on the Public Sphere" in *Habermas and the Public Sphere*, ed. Craig Calhoun (Cambridge: MIT Press, 1992), esp. pp. 436–39.

6. This interpretation evolved over several cases, notably *Bigelow v. Virginia*, 421 US 809 (1975); *Virginia State Board of Pharmacy v. Virginia Citizens Consumer Council, Inc.*, 425 US 748 (1976); and *First National Bank of Boston v. Bellotti*, 435 US 765 (1978).

7. Gerald Markowitz and David Rosner, *Deceit and Denial: The Deadly Politics of Industrial Pollution* (Berkeley: University of California Press, 2002).

8. Sven Birkerts, *The Gutenberg Elegies: The Fate of Reading in an Electronic Age* (New York: Fawcett Columbine, 1994).

Appendix

1. Martha Freeman, ed., *Always, Rachel: The Letters of Rachel Carson and Dorothy Freeman, 1952–1964* (Boston: Beacon, 1995).

2. Paul Brooks, *Two Park Street* (Boston: Houghton Mifflin, 1986).

3. Paul Brooks, *Speaking for Nature: How Literary Naturalists from Henry Thoreau to Rachel Carson Have Shaped America* (Boston: Houghton Mifflin, 1980).

4. Brooks, *Speaking for Nature*, 239.

5. Linda Lear, *Rachel Carson: Witness for Nature* (New York: Henry Holt, 1997). Lear has also edited a collection of previously unpublished works by Carson, *Lost Woods: The Discovered Writing of Rachel Carson* (Boston: Beacon, 1998). See also "Bombshell in Beltsville: The USDA and the Challenge of *Silent Spring*," *Agricultural History* 66 (Spring 1992): 151–70; and "Rachel Carson's *Silent Spring*," *Environmental History Review* 17 (Summer 1993): 23–48.

6. H. Patricia Hynes, *The Recurrent Silent Spring* (New York: Pergamon, 1989). See also Mary A. McCay, *Rachel Carson* (New York: Twayne, 1993); Carol Gartner's *Rachel Carson* (New York: Ungar, 1983); and for younger readers, Philip Sterling's *Sea and Earth: The Life of Rachel Carson* (New York: Thomas Y. Crowell, 1970). An e-book biography is now available on-line only through certain libraries: Liza N. Burby, *Rachel Carson: Writer and Environmentalist* (New York: Rosen Publishing Group, 1997), accessed at www.netlibrary .com/ebook_info.asp?product_id=19678.

7. Robert Darnton, "What Is the History of Books?" *Daedalus* 111 (Summer 1982): 67–69. This article also appears in Cathy N. Davidson's anthology, *Reading in America: Literature and Social History* (Baltimore: Johns Hopkins University Press, 1989).

8. See especially Adam Rome on the rise of environmentalism in the sixties in " 'Give Earth a Chance': The Environmental Movement and the Sixties," *Journal of American History* 90 (September 2003): 525–54. See also Edmund Russell, *War and Nature: Fighting Humans and Insects with Chemicals from World War I to Silent Spring* (Cambridge, Eng.: Cambridge University Press, 2001).

9. Rome's work is a good example and, interestingly, he cites many other books as agents of social change in the sixties. See also Thomas Dunlap's discussion placing *Silent Spring* in the context of public nervousness over negative effects of postwar science: *DDT: Scientists, Citizens, and Public Policy* (Princeton, N.J.: Princeton University Press: 1981).

10. James Whorton, *Before* Silent Spring: *Pesticides and Public Health in Pre-DDT America* (Princeton, NJ: Princeton University Press, 1974), viii. Whorton noted predecessor books: Upton Sinclair's *The Jungle;* Ruth deForest Lamb's *American Chamber of Horrors*; Arthur Kallett and F. J. Schlink's *100,000,000 Guinea Pigs*, an early 1930s sensationalized tract about poisonous residues of arsenic and lead sprays on food; and the 1936 follow-up, *40,000,000 Guinea Pig Children* on the special children's market evolving through heavy advertising.

11. Frank Graham, *Since Silent Spring* (Boston: Houghton Mifflin, 1970).

12. See, for example, M. Jimmie Killingsworth and Jacquelne Palmer on the "apocalyptic narratives" in environmental writing in "Millennial Ecology: The Apocalyptic Narrative from Silent Spring to Global Warming," in *Green Culture: Environmental Rhetoric in Contemporary America*, ed. Carl G. Herndl and Stuart C. Brown (Madison: University of Wisconsin Press, 1996), 21–45; and Scott Slovic's discussion of "jeremiadic" strategies in works like *Silent Spring* in "Politics in American Nature Writing," in Herndl and Brown, *Green Culture*, 101–5.

13. Craig Waddell, ed., *And No Birds Sing: Rhetorical Analyses of Rachel Carson's Silent Spring* (Carbondale: Southern Illinois University Press, 2000).

14. Valerie Jan Gunter, "Environmental Controversies, News Media, and the State: The Case of Synthetic Organic Pesticides in the 1940s, 1950s, and 1960s" (Ph.D. Diss., Michigan State University, 1995).

15. Andrew Jamison and Ron Eyerman, *Seeds of the Sixties* (Berkeley: University of California Press, 1994), 94.

16. Christopher J. Bosso, *Pesticides and Politics: The Life Cycle of a Public Issue* (Pittsburgh: University of Pittsburgh Press, 1987), 115.

17. Scott Slovic, *Seeking Awareness in American Nature Writing: Henry Thoreau, Annie Dillard, Edward Abbey, Wendell Berry, Barry Lopez* (Salt Lake City: University of Utah Press, 1992), 169.

18. Darnton, "What Is the History of Books?" 67–69. For an exceptionally concise and scholarly history of the field, see Joan Shelley Rubin, "What Is the History of the History of Books?" *Journal of American History* 90 (September 2003): 555–75.

19. Darnton, "What Is the History of Books?" 68.

20. John Sutherland, "Publishing History: A Hole at the Centre of Literary Sociology," *Critical Inquiry* 14 (Spring 1988): 574–75.

21. Elizabeth L. Eisenstein, *The Printing Revolution in Early Modern Europe*, vols. 1 and 2 (Cambridge: Cambridge University Press, 1983). This impressive work was condensed into the somewhat more readable *The Printing Press as an Agent of Change* (Cambridge, Eng.: Cambridge University Press, 1989).

22. Adrian Johns, *The Nature of the Book: Print and Knowledge in the Making* (Chicago: University of Chicago Press, 1998).

23. Scott E. Casper, Joanne D. Chaison, and Jeffrey D. Groves, eds., *Perspectives on American Book History: Artifacts and Commentary* (Amherst: University of Massachusetts Press, 2002).

24. John Tebbel, *A History of Book Publishing in the United States,* 4 vols. (New York: Bowker, 1981). His media history works include *The Media in America* (New York: Crowell, 1974) and, with Mary Ellen Zuckerman, *The Magazine in America 1741–1990* (New York: Oxford University Press, 1991)

25. Hellmut Lehmann-Haupt, *The Book in America: A History of the Making and Selling of Books in the United States* (New York: Bowker, 1951); Lewis A. Coser, Charles Kadushin, and Walter W. Powell, *Books: The Culture and Commerce of Publishing* (New York: Basic, 1982).

26. Orion H. Cheney, *Economic Survey of the Book Industry, 1930–31* (New York: National Association of Book Publishers, 1931).

27. Characteristically written by former editors or publishers, they include works like Charles Madison, *Book Publishing in America* (New York: McGraw-Hill, 1966); Chandler Grannis, *What Happens in Publishing* (New York: Crown, 1967); John Dessauer, *Book Publishing,* (New York: Continuum, 1989); Howard Greenfeld, *Books from Writer to Reader* (New York: Crown, 1976); and Clarkson Potter, *Who Does What and Why in Book Publishing* (New York: Birch Lange, 1990).

28. For example, Elizabeth Geiser's anthology, *The Business of Book Publishing: Papers by*

Practitioners (Boulder, Colo.: Westview, 1985), offers an overview of various editorial, production, and marketing aspects of the publishing process.

29. Virginia Woolf, "Reviewing," in *Collected Essays*, vol. 2 (New York: Harcourt, Brace & World, 1967), 205.

30. Woolf, "Reviewing," 204.

31. Granville Hicks, "The Journalism of Book Reviewing," *Saturday Review*, 12 December 1959, 16.

32. Clifton Fadiman, "The Reviewing Business," *Gentlemen, Scholars and Scoundrels: A Treasury of the Best of Harper's Magazine from 1850 to the Present*, edited by Horace Knowles (New York: Harper & Brothers, 1959); John E. Drewry, *Writing Book Reviews* (Westport, Conn.: Greenwood Press, 1966), 11.

33. Joan Shelley Rubin, *The Making of Middlebrow Culture* (Chapel Hill: University of North Carolina Press, 1992), 35–37; Robert Kirsch, "The Importance of Book Reviewing," in *Book Reviewing: A Guide to Writing Book Reviews for Newspapers, Magazines, . . .* ed. Sylvia E. Kamerman (Boston: The Writer, 1978), 3.

34. Frank Donoghue discusses reviewers' ability to confer celebrity in *The Fame Machine: Book Reviewing and Eighteenth-Century Literary Careers* (Stanford: Stanford University Press, 1996). Carlin Romano made the highly debatable complaint that books are not reported by the press as news, that coverage of books only takes place in reviews in "Extra! Extra! The Sad Story of Books as News," *Media Studies Journal* 6 (summer 1992): 123–32. See also Victoria Glendenning, "The Book Reviewer: The Last Amateur?" in *Essays by Divers Hands*, ed. A. N. Wilson (Woodbridge, Suffolk, Eng.: Royal Society of Literature, 1986).

35. See recent examples, particularly Gerry O'Sullivan, "Against the Grain: The Politics of Book Reviewing," *Humanist* (November/December 1990): 43–44; and Steve Weinberg, "The Unruly World of Book Reviews," *Columbia Journalism Review* 28 (March/April 1990): 51–54. For other examples, see "Book Reviewing in America: A Forum," *Review of Contemporary Fiction* 9 (Spring 1989): 216–35; Gerald Howard, "The Cultural Ecology of Book Reviewing," *Media Studies Journal* 6 (Summer 1992): 90–110; and Bill Marx, "The Decline of Book Reviewing," *Publishers Weekly* 240 (25 October 1993): 36.

36. J. Bowman, "A Little Help for Their Friends," *National Review*, 7 March 1994, 63.

37. O'Sullivan, "Against the Grain," 43–44.

38. Pat Rogers, "Just a Hint of Scandal: The Birth of Book Reviewing," *New York Times Book Review*, 1 January 1984, 16.

39. Coser, Kadushin, and Powell, *Culture and Commerce of Publishing*, 308–32.

40. Ann B. Haugland, "Defining the Book: Culture and Commerce in the *New York Times Book Review*" (PhD. Diss., University of Iowa, 1992). See also Bowman, "A Little Help for Their Friends," 63–67; Francis Brown, *The Story of the New York Times Book Review* (New York: New York Times Book Review, 1969); Scott Walker, "A Review in the Times?! Oh, No!" *Publishers Weekly* 240 (11 October 1993): 45.

41. Rubin, *Making of Middlebrow Culture*, 41.

42. R. S. Fogerty, editorial, *Antioch Review* 49 (fall 1991): 483.

43. Nina Baym, *Novels, Readers, and Reviewers: Responses to Fiction in Antebellum America* (Ithaca: Cornell University Press, 1984), 21.

44. Drewry, *Writing Book Reviews*, 11.

45. See Roger Chartier, *The Order of Books: Reader, Authors, and Libraries in Europe between the Fourteenth and Eighteenth Centuries* (Stanford: Stanford University Press, 1994), 1–24; and note Stanley Fish's "interpretive communities" in *Is There a Text in This Class? The Authority of Interpretive Communities* (Cambridge: Harvard University Press, 1980), as well as Benedict Anderson's "imagined communities" in *Imagined Communities* (London: Verso, 1983, 1991).

46. See David D. Hall, *Worlds of Wonder, Days of Judgment: Popular Religious Belief in Early New England* (Cambridge: Harvard University Press, 1989); Michael Warner, *The Letters of the Republic: Publication and the Public Sphere in Eighteenth-Century America,* (Cambridge: Harvard University Press, 1990); Richard H. Brodhead, *Cultures of Letters: Scenes of Reading and Writing in Nineteenth Century America* (Chicago: University of Chicago Press, 1993); and also Anderson, *Imagined Communities.*

47. See particularly Davidson's anthology, *Reading in America,* and her *Revolution and the Word: The Rise of the Novel in America* (New York: Oxford, 1986).

48. In addition to Rubin's *Making of Middlebrow Culture,* see Lawrence Levine, *High-brow/Lowbrow: The Emergence of Cultural Hierarchy in America* (Cambridge: Harvard University Press, 1988), and Michael Denning's exploration of the dime novel and its role in the lives of working-class male readers in *Mechanic Accents* (London: Verso, 1987).

49. Joan Rubin's chapter in *Making of Middlebrow Culture* is "Why Do You Disappoint Yourself?" See also Janice A. Radway, "The Book-of-the-Month Club and the General Reader: The Uses of 'Serious Fiction,' " in Davidson, *Reading in America,* 259–84; Janice A. Radway, *Reading the Romance: Women, Patriarchy, and Popular Literature* (Chapel Hill: University of North Carolina Press, 1984).

50. Robert Darnton, *Forbidden Best-Sellers of Pre-Revolutionary France* (New York: W. W. Norton, 1995), 77.

51. *New York Public Library's Books of the Century,* edited by Elizabeth Diefendorf (New York: Oxford University Press, 1996). A few other examples are: Malcolm Cowley and Bernard Smith, eds., *Books That Changed Our Minds* (New York: Doubleday, Doran, 1939); Robert B. Downs, *Books That Changed the World* (New York: Mentor, 1956); Robert Downs, *Books That Changed America* (New York: Macmillan, 1970). Alice Payne Hackett's annual series on the top ten best-sellers was published intermittently between 1950 and 1975; see also Alice Payne Hackett and James Henry Burke, *80 Years of Best Sellers: 1895–1975* (New York: Bowker, 1978).

52. Frank Luther Mott, *Golden Multitudes* (New York: Macmillan, 1947).

53. Peter Mann, *From Author to Reader: A Social Study of Books* (London: Routledge & Kegan Paul, 1982), 20.

54. See Darnton's elaborate schema of "The Communication Circuit-News" in *Forbidden Best-Sellers,* 189; Thomas Whiteside, *The Blockbuster Complex: Conglomerates, Show Business, and Book Publishing* (Middletown, Conn.: Wesleyan University Press, 1981).

55. See, for example, David Hall's *Worlds of Wonder* or Darnton's "Readers Respond to Rousseau," in *The Great Cat Massacre and Other Episodes in French Cultural History* (New York: Vintage, 1985).

56. David Paul Nord, "Reading the Newspaper: Strategies and Politics of Reader Response, Chicago, 1912–1917," *Journal of Communication* 45 (Summer 1995): 66–93. Also see Brian Thornton, " 'Gospel of Fearlessness' or 'Outright Lies': A Historical Examination of Magazine Letters to the Editor, 1902–1912 and 1982–1992," *American Journalism* 15 (spring 1998): 37–57; "The Moon Hoax: Debates about Journalistic Ethics in Four New York Newspapers in 1835," *Journal of Mass Media Ethics* 15 (Fall 2000): 89–100; "The Disappearing Media Ethics Debate in Letters to the Editor," *Journal of Mass Media Ethics* 13 (Fall 1998): 40–55; and "Telling It Like It Is: Letters To The Editor Discuss Journalism Ethics in 10 American Magazines, 1962, 1972, 1982 and 1992," *Journal of Magazine and New Media Research* 1 (Spring 1999), accessed on-line at http://www.bsu.edu/web/aejmcmagazine/journal/archive/default.html.

57. David Thelen, *Becoming Citizens in the Age of Television: How Americans Challenged the Media and Seized Political Initiative During the Iran-Contra Debate* (Chicago: University of Chicago Press, 1996); and Karin Wahl-Jorgensen, *Constructing the Public: Letters to the Editor as a Forum for Deliberation* (Diss., Stanford University, 2000). Also see Wahl-

Jorgensen's "Letters to the Editor as a Forum for Public Deliberation: Modes of Publicity and Democratic Debate," *Critical Studies in Media Communication* 18 (September 2001): 303–20; "Letters to the Editor, the Public Sphere, and the Metaphor of the Marketplace," *Peace Review* 11 (March 1999): 53–59; "The Normative-Economic Justification for Public Discourse: Letters to the Editor as a 'Wide Open' Forum," *Journalism and Mass Communication Quarterly* 79 (Spring 2002): 121–33.

58. See Walter Lippmann, *Public Opinion* (New York: Macmillan, 1922); Harold D. Lasswell, "The Structure and Function of Communication in Society," in Lyman Bryson, ed., *Communication of Ideas: A Series of Addresses* (New York: Institute for Religious and Social Studies, 1948), 37–52; and Paul F. Lazarsfeld and Robert K. Merton, "Mass Communication, Popular Taste and Organized Social Action," in Bryson, *Communication of Ideas,* 95–118.

59. Bernard C. Cohen, *The Press and Foreign Policy* (Princeton, N.J.: Princeton University Press, 1963), 13.

60. Maxwell McCombs and Donald L. Shaw, "The Agenda-Setting Function of the Mass Media," *Public Opinion Quarterly* 36 (1972): 176–87.

61. Whether the media actually do set the agenda of public discussion is a question that has of course been investigated, especially in light of a possible "post hoc" fallacy in agenda-setting research, i.e., whether rising public attention that seems to follow media coverage is absolute evidence of causality. See Everett M. Rogers and James W. Dearing, "Agenda-Setting Research: Where Has It Been, Where Is It Going?" in *Communication Yearbook,* vol. 11, ed. J. A. Anderson, (Newbury Park, Calif.: Sage, 1988), 555–94.

62. Rogers and Dearing, "Agenda-Setting Research," 565.

63. See Donald L. Shaw, "The Press Agenda in a Community Setting," in D. L. Shaw and M. E. McCombs, eds., *The Emergence of American Public Issues: The Agenda-Setting Function of the Press* (St. Paul, Minn.: West, 1977), 19–31.

64. Lippmann, *Public Opinion,* 9.

65. Lang and Lang write: "In fact, issues have been variously conceptualized as (1) *concerns,* the things about which people are personally worried; (2) *perceptions of key problems* facing the country, about which the government should do something; (3) the existence of *policy alternatives* between which people must choose . . . ; (4) a *public controversy,* such as the one over Watergate; and (5) the "reasons" or underlying determinants of *political cleavage. . . .*" Gladys Engel Lang and Kurt Lang, "Watergate: An Exploration of the Agenda-Building Process," *Mass Communication Review Yearbook* 2 (Newbury Park, Calif.; Sage, 1981), 450 (original emphasis).

66. G. Ray Funkhouser, "The Issues of the Sixties: An Exploratory Study in the Dynamics of Public Opinion," *Public Opinion Quarterly* 37 (1973): 62–75; also in Funkhouser's "Trends in Media Coverage of the Issues of the '60s," *Journalism Quarterly* 50 (1973): 533–38. Daniel Boorstin coined the term "pseudo-events" for such occasions as Earth Day.

67. Anthony Downs, "Up and Down With Ecology: The 'Issue-Attention Cycle,' " *The Public Interest* (Spring 1972): 43.

68. Lang and Lang, "Watergate," 465. Also see David Weaver and Swanzy Nimby Elliot, "Who Sets the Agenda for the Media? A Study of Local Agenda-Building." *Journalism Quarterly* 62 (Spring 1985): 87–94.

69. See, among others, Marshall McLuhan's *Understanding Media* (New York: Signet, 1964).

70. John A. Fortunato raised the issue of the role of public-relations departments in setting the media's agenda in "Public Relations Strategies for Creating Mass Media Content: A Case Study of the National Basketball Association," *Public Relations Review* 26 (Winter 2000): 481–97. Kyle Huckings examined the success of the "Christian coalition"

in affecting general media coverage of the group; "Interest-Group Influence on the Media Agenda: A Case Study," *Journalism and Mass Communication Quarterly* 76 (spring 1999): 76–86.

71. "The events which are not scored are reported either as personal and conventional opinions, or they are not news. They do not take shape until somebody protests, or somebody investigates, or somebody publicly, in the etymological meaning of the word, makes an *issue* of them"; Lippmann, *Public Opinion,* 10 (original emphasis).

72. Funkhouser, "Issues of the Sixties," 71–72.

73. Bruce H. Westley, "What Makes It Change?" *Journal of Communication* 26 (Spring 1976): 46–47.

74. Rogers and Dearing, "Agenda-Setting Research," 558.

75. Lucig H. Danielian and Stephen D. Reese, "A Closer Look at Intermedia Influences on Agenda Setting: The Cocaine Issue of 1986," in *Communication Campaigns about Drugs: Government, Media, and the Public,* ed. Pamela J. Shoemaker, (Hillsdale, N.J.: Lawrence Erlbaum, 1989), 47–66.

76. Oscar H. Gandy Jr., *Beyond Agenda Setting: Information Subsidies and Public Policy* (Norwood, N.J.: Ablex, 1982), 8.

77. "Sources enter into an exchange of value with journalists in which they (1) they reduce the costs of news work to increase their control over news content; (2) they reduce the costs of scientific research to increase their control over scientific and technical information; and (3) they even reduce the costs of writing and producing television fiction to increase their control over the cultural background against which social policy questions are generally framed." Gandy, *Beyond Agenda Setting,* 15.

78. Philip Meyer, "Accountability When Books Make News," *Media Studies Journal* (Summer 1992): 136–37.

79. Everett M. Rogers with F. Floyd Shoemaker, *Communication of Innovations* (New York: Free Press, 1971); and Elihu Katz and Paul F. Lazarsfeld, *Personal Influence: The Part Played by People in the Flow of Mass Communications* (New York: Free Press, 1964).

80. Examples of this approach include: Anthony Downs's study of the "issue-attention" cycle in "Up and Down With Ecology"; Funkhouser's studies of 1960s media and public opinion dynamics in "Issues" and "Trends"; Danielian and Reese's studies on the media and the drug issue in "A Closer Look" and "Intermedia Influence and the Drug Issue: Converging on Cocaine" in Shoemaker, *Communication Campaigns,* 29–46; and Fay Lomax Cook and Wesley Skogan's study of convergence in the "life course" of policy issues in "Convergent and Divergent Voice Models of the Rise and Fall of Policy Issues," in *Agenda Setting: Readings on Media, Public Opinion, and Policymaking,* ed. David L. Protess and Maxwell McCombs (Hillsdale, N.J.: Lawrence Erlbaum, 1991).

81. Funkhouser, "Trends," 534–36.

82. Rogers and Dearing arrayed these three agendas (media, public, and policy) side by side, noting the three impact vectors of "gatekeepers, influential media, and spectacular news events," "personal experience and interpersonal communication among elites and other individuals," and "realworld indicators of the importance of an agenda issue or event." Rogers and Dearing, "Agenda-Setting Research," 557. Jarol Manheim took the same triad of agendas and arrayed them each in a three-dimensional model to allow for gradations in visibility, salience, and valence (positive or negative favor); the three agendas thus schematized formed a model in which they influence or interact with one another. Jarol B. Manheim, "A Model of Agenda Dynamics," in *Communication Yearbook,* vol. 10, ed. Margaret McGlaughlin (Newbury Park, Calif.: Sage, 1987), 499–516.

83. Danielian and Reese, "Closer Look."

84. Klaus Schoenbach, "Agenda-Setting Effects of Print and Television in West Germany," in Protess and McCombs, *Agenda Setting,* 129.

85. Marc Benton and P. Jean Frazier, "The Agenda-Setting Function of the Mass Media at Three Levels of 'Information Holding'," *Communication Research* 3 (1976): 261–74.

86. For examples of studies in this area, see Elihu Katz, "Mass Communication Research and the Study of Popular Culture," *Studies in Public Communication* 2 (1959): 1–6; Elihu Katz, Jay G. Blumler, and Michael Gurevitch, "Utilization of Mass Communication by the Individual," in *The Uses of Mass Communications: Current Perspectives on Gratifications Research,* ed. J. G. Blumler and E. Katz (Beverly Hills, Calif.: Sage, 1974); Sandra J. Ball-Rokeach and Melvin L. DeFleur, "A Dependency Model of Mass-Media Effects," *Communication Research* 3 (1976): 3–21; and Sandra J. Ball-Rokeach, "The Origins of Individual Media-System Dependency: A Sociological Framework," *Communication Research* 12 (1985): 485–510.

87. Radway, *Reading the Romance.*

88. Elihu Katz, Michael Gurevitch, and Hadassah Haas, "On the Use of the Mass Media for Important Things," *American Sociological Review* 30, no. 2 (1973): 164–81. They found that books were more similar to newspapers and cinema than to the broadcast media in their ability to help meet informational and diversion needs.

89. See, for example, discussion of changes in the experience and concept of the book form in Sven Birkerts *The Gutenberg Elegies: The Fate of Reading in an Electronic Age* (New York: Fawcett Columbine, 1994).

90. Jürgen Habermas, "The Public Sphere," in *Rethinking Popular Culture: Contemporary Perspectives in Cultural Studies,* ed. Chandra Mukerji and Michael Schudson (Berkeley: University of California Press, 1991), 398. Habermas originally wrote *The Structural Transformation of the Public Sphere: An Inquiry into a Category of Bourgeois Society* in 1962, and MIT Press published an English translation in 1989 (current paperback edition, Cambridge: MIT Press, 2001). See also Jürgen Habermas, "Further Reflections on the Public Sphere," in *Habermas and the Public Sphere,* ed. Craig Calhoun (Cambridge: MIT Press, 1992), 452–55, which represents his own reflections, thirty years later, on the original work and his response to various critiques of it.

91. Calhoun, *Habermas,* 23; also see Calhoun's introduction for a general discussion of the concept, in *Habermas,* 1–48. Extensive discussions can be found in Habermas's *Structural Transformation* and *Communication and the Evolution of Society* (Boston: Beacon, 1979).

92. Calhoun, introduction, *Habermas,* 6.

93. Jason Epstein, "A Criticism of Commercial Publishing," *The American Reading Public: What It Reads, Why It Reads* (New York: Bowker, 1963).

94. Michael Schudson, *The Power of the News* (Cambridge: Harvard University Press, 1995), 197.

95. Schudson, *Power of News,* 19.

96. Schudson, *Power of News,* 25.

97. Calhoun, "Introduction," *Habermas,* 37.

98. See particularly Habermas, *Structural Transformation,* 194–95, on the significance of public relations.

99. Habermas, "Further Reflections," in Calhoun, *Habermas,* 455.

100. Habermas, "Further Reflections, in Calhoun, *Habermas,* 452.

INDEX

Note: Page references to illustrations are in *italics.*

ABC News, 131, 150
advertising, role of, 51, 65, *78, 82*, 111, 115–16, *152*, 178
agenda setting, 127, 156–57, 185, 212–17, 219, 249n. 82
agenda building, 213, 215
agents, literary, 28–29
Agricultural Research Service (ARS), 91–92, 138
Agriculture, Department of. *See* U.S. Department of Agriculture
American Cyanamid, 141, 167
American Medical Association (AMA), 111, 146–47
Associated Press, 84
Atkinson, Brooks, 155
Atlantic Monthly, 147–48
Audubon Magazine, 15, 145
Audubon Society. *See* National Audubon Society

balance of nature, 8, 30, 105
Baldwin, I. L., 91, 101, 103, 112, 143–44, 146
Bates, Marston, 147
Barnes, Irston, 135
BeechNut, 64
Benson, Ezra Taft, 44–45, 106
Benton, Marc, 216
Better Homes and Gardens, 146
book: definition of, 183–85, 192–93, 218; form, 17–18, 48–52, 120–27, 177–79, 181–82, 183–84, 192–3; future of, 192–98; history, 204–11; vs. magazine, 9–10, 48–52, 85, *86,* 108–10,

155–56, 177–79; publishing industry, 205–6; "unread," 50, 118, 156–58, *157,* 179. *See also* books' function; reviewing, book
Book-of-the-Month Club, 4, 26, 39, 73, 74, 209–10
books' function, 2–3, 89, 117–18, 184–98; for author, 48–52; as information subsidy, 186, 215, 249n. 77; for media, 119–27; for opposition, 89, 117–18; for publishers, 81–84; for society, 12, 117–18, 184–98, 209–11
Boothbay Register, 43, 47
Bosso, Christopher, 203
Boston Globe, 76, 114, 135
Bourdieu, Pierre, 120, 205
Boyer, Paul, 12–13
Brinkley, Parke, 104, 108–9, 113, 154, 161
Brodhead, Richard, 209
Brooks, Paul, xiii, 3, 7, 24, 37, 59–62, *60,* 66–81, 88, 199, 200
Buchheister, Carl, 68, 145
Business Week, 96, 98, 144

Calhoun, Craig, 220
Canadian Film Board, 150–51
Canham, Erwin, 150
Carleton, R. Milton, 47, 114
Carson, Rachel, *21, 42, 142;* and audience, 36–39, 179–82; biography, 5, 20–29, 228n. 78; health, 5, 27, 44, 228n. 66; and interviews, 122–24; and media, 26–28, 39–48, 124, *125;* mission, 3–5, 29–36, 179–82; obituaries, 154; work style, *22,* 25–28
CBS, 5, 24, 41–43, 80, 115–16

CBS Reports, 12, 24, 41–43, 80, 105, 106, 122, 124, 126–27, 130, 150, 151–55; advertisers for, 116, 237n. 109; "The Silent Spring of Rachel Carson," 5, 17, 24, 80, 105, 106, 115–16, 122, 124, 126–27, 130, 150, *152;* "The Verdict on Rachel Carson's Silent Spring," 31, 153–54

Chartier, Roger 209

chemical industry, 141, 234n. 16. *See also* Manufacturing Chemists Association; National Agricultural Chemicals Association; *and individual companies*

Chemical and Engineering News, 91, 98, 99, 101, 149, 162–63, 174

Cheney, O. H., 205

Chicago Tribune, 136, 155–56

Christian Science Monitor, 73, 76, 150

Clement, Roland, 91, 161–62

Cohen, Bernard, 212

Commonweal, 147

"Communications Circuit." *See* Darnton, Robert

congressional hearings. *See* U.S. Senate hearings

conservation. *See* environmentalism

Consumer Reports, 70, 145–46

Consumers Union, 4, 70, 145–46

Cornell Report. See *Facts on the Use of Pesticides*

Coser, Lewis, 205, 208

Cottam, Clarence, 42, 91

cranberry scare, 10–11, 25, 150–51

Danielian, Lucig, 214–16

Darby, William, J. 85, 101, 104, 162

Darnton, Robert, 200–202, 204, 210; "Communications Circuit," 200, *201*

Davidson, Cathy, 209

DDT, 14–15, 23, 38, 131, 141, 150, 154, 159

Dearing, James, 212–13, 214

Decker, George C., 91, 95, 101, 146

Des Moines Register, 15, 131, *132*, 136, 163

"Desolate Year" (Monsanto), 95, 100, 105, 111, 113, 115, 121, 129

Diamond, Edwin, 24, 26, 42, 51, 142–43, 162, 285n. 6

direct mail, 111–12, 113

Douglas, William O., 15, 26, 46, 67, 121

Dow Chemical, 104, 141

Downs, Anthony, 213

Drewry, John, 207, 209

DuPont, 71–72, 97, 98, 109, 110, 141

ecology. *See* environmentalism

Edge of the Sea (Carson), 3, 23, 40, 50, 57, 61, 62

Eiseley, Loren, 148

Eisenstein, Elizabeth, 204

environmentalism, 13–16, 124, 144, 147, 154, 155, 191, 200, 202, 213, 225n. 21,

Epstein, Jason, 220

eradication programs, 7, 15, 16, 33, 90–91, 96, 227n. 45

events vs. issues, and media coverage, 117, 119–20, 212–15

Eyerman, Ron, 203

Fact and Fancy (NACA), 48, 78–79, 85, 100, 110, 111, 113

Facts on the Use of Pesticides (Cornell Report), 101

Fadiman, Clifton, 207

fallout. *See* radioactive fallout

farmers and farming, 11, 131–32

Farm Journal, 144

Farm Magazine, 144–45

Field & Stream, 145

Fire Ant on Trial (film), 16, 35

fire-ants. *See* eradication programs

Fish and Wildlife Service. *See* U.S. Fish and Wildlife Service

Food and Drug Administration (FDA), 7, 34, 93–94

Ford, Anne, 29, 62, 71, 73, 74, 75, 76–77, 84–85, 107, 114, 135

Forrestal, Dan, 95, 108

Frazier, P. Jean, 216

Free Deeds (magazine), 174, 242n. 34

Freeman, Dorothy, 28, 30, 199, 226n. 17

Freeman, Orville, 92, 153, 176

Friendly, Fred, 116

Funkhouser, G. Ray, 213, 214, 216

Galton, Lawrence, 134

Gandy, Oscar, 186, 215

Good Housekeeping, 146

Graham, Frank, 202

Greenstein, Milton, 64–65

Gunter, Valeria Jan, 202

Gurevitch, Michael, 217

Haas, Hadassah, 217

Habermas, Jürgen, 127, 155–56, 194, 215, 219–21, 225n. 29, 250n. 90

Hall, David, 209

Hansen, Harry, 155–56, 208

Harrison, E. Bruce, 94–95

Hersey, John, 55–56, 57

Hicks, Granville, 207

Hiroshima (Hersey), in *New Yorker*, 55–56, 57

History of the Book in America (Cambridge University Press series), 204–5

Horgan, Thomas, 84

Houghton Mifflin, 57–62, 66–81, 85–88; history of, 57–62, 230n. 28; and publicity, 43, 44, 69–75, 77–80; relation to *New Yorker*,

81–85; rights sales, 70–71, 73; *Story of "Silent Spring,"* 79–80
Huckins, Olga, 38, 159, 227n. 55
Hynes, Patricia, 200

independence, 185–86, 191, 194–95, 198, 220–21; of author, 50–52 ; perception of, 181, 188–89; of publishers, 56–57, 65, 86–88
information subsidy, 186, 215, 249n. 77
issues vs. events, and media coverage, 117, 119–20, 212–15

Jamison, Andrew, 203
Johns, Adrian, 204
Jukes, Thomas, 95, 107, 114, 167
Jungle (Sinclair), 189, 190, 198

Kadushin, Charles, 205, 208
Katz, Elihu, 215, 217
Kennedy, John F., 4, 24
King, C. G., 96, 104, 107, 148
Kirsch, Robert, 207

Lang, Gladys Engel, 213
Lang, Kurt, 213
Larrick, George, 153
Lasswell, Harold, 212
Laycock, Ted, 114, 135
Lazarsfeld, Paul, 212, 215
Lear, John, 148
Lear, Linda, 28, 51, 200
Lederer, William J., 164–65
Lehmann-Haupt, Hellmut, 205
Leonard, Rodney, 92, 109
letters, 38, 93, 156–64, 168–73, 211; to the editor, 38, 160–64, 181, 241n. 3; evidence of action in, 93, 173–77, 179–82; function of, 159–64, 164–77, 181, 211; to the *New Yorker*, 64, 166–68, 169–72
Levine, Laurence, 209
Life, 32, 40, 51, 141–42, *142*, 161–62
Lippmann, Walter, 212, 214
Long Island lawsuit, 3, 15, 23, 26, 30, 66, 67, 133, 139, 148
Los Angeles Times, 136

magazines. *See* media
Magnuson committee. *See* U.S. Senate hearings
Mann, Peter, 211
Manchester, Harlan, 114
Manufacturing Chemists Association (MCA), 94–96, 99, 113–14, 116
marketplace of ideas, 156–57, 220
McCombs, Maxwell, 212
McLean, Louis, 4, 5, 64–65, 72–73, 100–101, 107–8. See also *Necessity, Value, and Safety of Pesticides*

McLuhan, Marshall, 214
McMullen, Jay 24, 115–16, 124, 153
media, role of, 127–33, 202–4; newspapers, 133–36, *152*; magazines, 136–49, 239n. 52; radio, 77, 114, 115, 124, 149–50; television, 150–55. See also Carson, Rachel: and media; *CBS Reports; individual magazines and newspapers*
Merton, Robert, 212
Meyer, Agnes, 26
Meyer, Philip, 2, 215
Monsanto, 95, 99–100, 105, 121. *See also* "Desolate Year"
Moore, Ernest, G. 92, 114
Mott, Frank Luther, 210

NAC News, 95
Nation, 147
Nation of Sheep (Lederer), 164–65
National Academy of Sciences–National Research Council (NAS-NRC), 91, 94, 101, 102, 106, 109, 110, 134
National Agricultural Chemicals Association (NACA), 48, 77–79, 85, 99, 100, 103, 104, 112, 113, 116, 154, 161
National Audubon Society, 15, 25, 60, 61, 68, 91, 135, 145
National Educational Television (NET), 150–51
Natural History, 144
Necessity, Value, and Safety of Pesticides (Velsicol), 101, 110
New Republic, 148–49
newspapers. *See* media
Newsweek, 139
New York Herald-Tribune, 76, 134–35
New York Times, 15–16, 63, 105, 115, 133–34, 153, 155, 163, 165–6, 208, 224n. 2
New York Times Book Review, 113, 134, 174
New York Times Magazine, 134
New Yorker: history of, 23–24, 50–51, 53–57; letters to, 64, 166–68, 169–72; publication of article in, 3–5, 9–10, 62–66, 85–88; relation to Houghton Mifflin, 3–5, 81–85
Nord, David Paul, 211
nuclear fallout. *See* radioactive fallout
Nutrition Foundation, 96, 107, 116

Ochs, Adolph 207

pesticides, discussion of, 14–16, 17–18, 96–97, 234n. 16
Peterson, Roger Tory, 26, 161
"Poisons, Pests, and People" (tv program), 150–51
Powell, Walter, 205, 208
Prevention, 147
PR News, 105, 111

Printers' Ink, 96, 97, 99, 105, 109, 111
President's Science Advisory Committee
    (PSAC), 4, 17, 24, 34, 48, 75, 80–81, 93–94,
    96, 126–27, 138–39, 153–54
public, concept of, 127–28, 218–20
Public Health Service. *See* U.S. Public Health
    Service
public sphere, 127–28, 143, 156–57, 159–60,
    173, 193–95, 219–21, 225n. 29

radiation. *See* radioactive fallout
radio. *See* media
radioactive fallout, 10, 12–13, 66–67, 164
Radway, Janice, 209–20, 217
Rand, J. H., 175
reviewing, book, 113, 122, *123*, 128–29, 149, 158,
    206–9; reader response and reading, 209–11
Reader's Digest, 40, 114, 140–41
Reese, Stephen, 214m 216
Ribicoff committee. *See* U.S. Senate hearings
Rodell, Marie, 29, 23, 24, 27, 28–29, 31, 37, 40–
    48, 68, 71–72, 141, 167
Rogers, Everett, 212–213, 214, 215
Ross, Harold, 54–56, 138, 239n. 41
Rostand, Jean, 6, 20, 30, 180
Rubin, Joan Shelley, 207, 208–9

Saturday Evening Post, 15–16, 113, 142–43, 161–
    62
Saturday Review of Literature, 15, 148
Schoenbach, Klaus, 216
Schudson, Michael, 220
Schweitzer, Albert, 30–31
Science, 91, 101, 143–44
Scientific American, 144
Sea Around Us (Carson), 23, 26, 28, 30, 51, 113
Senate hearings. *See* U.S. Senate hearings
Sevareid, Eric, 41, *42*, 130, 136, 150
Shaw, Byron T., 92–93, 112, 135, 138–39
Shaw, Donald, 212
Shawn, William, 23, 24, 37, 54–57, 62–66, 81–
    82
Siegel, Kalman, 166, 181, 241n. 3
Silent Spring: publication history, 3–5, 23–25,
    177; research and documentation ("fifty-
    five pages"), 4, 8, 45–46, 49–50, 85–86, 109,
    120, 172–73; synopsis, 5–9; title, 7, 67, 68
Slovic, Scott, 203–4, 206–7
Sports Illustrated, 145, 161–62
Stare, Frederick 96, 101, 103–4, 111, 112, 124,
    134–35, 146, 172–73
Story of "Silent Spring" (Houghton Mifflin),
    79–80

Strother, Robert, 140
Sullivan, Walter, 115, 134
Supreme Court. *See* Long Island lawsuit
systems theory, 213–14

Tebbel, John, 205
television. *See* media
thalidomide, 37, 102, 148, 164, 203
Thelen, David, 211
Thompson, Lovell, 29, 59, 61–62, 69, 74, 85,
    87–88
Thornton, Brian, 211
Time, 113, 137–38, 146, 154, 161–62
Today's Health, 76, 146
True, 145
TV Guide, 151, *151*

Udall, Stewart, 26, 61, 145, 161
Uncle Tom's Cabin (Stowe), 134, 152, 189, 190,
    198
"unread book." *See* book: "unread"
U.S. Department of Agriculture (USDA), 4–
    5, 16, 32–33, 35, 73, 90–94, 112, 114–15, 175–
    76
U.S. Fish and Wildlife Service (FWS), 15, 20.
    32–33, 93–94, 153
U. S. News & World Report, 93, 135, 138–39
U.S. Public Health Service, 93–94
U.S. Senate hearings, 17, 24–25, 31, 34, 47
"uses and gratifications" theory, 217

Van Fleet, Clark, 148
Velsicol, 4, 5, 72–73, 100–101, 107–8, 141

Waddell, Craig, 202
Wahl-Jorgensen, Karen, 211
Warner, Michael, 201
Washington Post, 26, 43, 135, 154
WEEI (Boston radio station), 77
Westcott, Cynthia 106–7, 228n. 87
Westley, Bruce 214
WHDH (Boston tv station), 150
White, E. B., 3, 23, 30
Whiteside, Thomas, 211
White-Stevens, Robert, 41, 104–5, 106, 112,
    124, 126, 151–57
Whorton, James, 202
Wiesner, Jerome, 92, 94, 240n. 68
Woolf, Virginia, 207
World Health Organization (WHO), 47, 131,
    144, 235n. 44

Yankee Magazine, frontis.